西餐烹调工艺

余松筠 主 编
丁 辉 李 俊 副主编

华中科技大学出版社
http://www.hustp.com
中国·武汉

内 容 简 介

本书的编写体现了简洁性、实用性和创新性等特点,秉持学以致用的理念,将西餐烹调工艺的基本知识与烹饪实践紧密结合,突出理论知识的应用性和对实践的指导性。

本书为烹饪与营养教育、食品质量与安全、西餐工艺等专业西餐课程的教学用书,也可作为西餐爱好者的学习参考书。全书涵盖了西餐烹调工艺的基本理论知识及常规实训项目等内容。

图书在版编目(CIP)数据

西餐烹调工艺/余松筠主编.—武汉:华中科技大学出版社,2018.8(2024.1重印)
ISBN 978-7-5680-4191-1

Ⅰ.①西… Ⅱ.①余… Ⅲ.①西式菜肴-烹饪 Ⅳ.①TS972.118

中国版本图书馆 CIP 数据核字(2018)第 182042 号

西餐烹调工艺 余松筠 主编
Xican Pengtiao Gongyi

策划编辑:汪飒婷
责任编辑:毛晶晶
封面设计:刘 婷
责任校对:曾 婷
责任监印:周治超

出版发行:华中科技大学出版社(中国·武汉)　　电话:(027)81321913
　　　　　武汉市东湖新技术开发区华工科技园　　邮编:430223
录　　排:华中科技大学惠友文印中心
印　　刷:广东虎彩云印刷有限公司
开　　本:787mm×1092mm　1/16
印　　张:13　插页:3
字　　数:283千字
版　　次:2024年1月第1版第5次印刷
定　　价:58.00元

本书若有印装质量问题,请向出版社营销中心调换
全国免费服务热线:400-6679-118　竭诚为您服务
版权所有　侵权必究

Foreword 前 言

本书为烹饪与营养教育、食品质量与安全、西餐工艺等专业西餐课程的教学用书,也可作为西餐爱好者的学习参考书。全书涵盖了西餐烹调工艺的基本理论知识及常规实训项目等内容。

本书对西餐烹饪专业知识体系进行规划、整合,提炼知识要素,优化教材内容,以提高学生的专业素养、实践动手能力,培养学生的创新精神为目标。教材内容按照理论实践一体化的教学模式的要求,以实践模块的思路进行编写。

本书的编写体现了简洁性、实用性和创新性等特点,秉持学以致用的理念,将西餐烹调工艺的基本知识与烹饪实践紧密结合,突出理论知识的应用性和对实践的指导性。

与国内同类教材比较,本书具有以下特点。

第一,内容的划分体现由浅入深、由简单到复杂的特点,层层递进,符合学生的认知规律。先理论,后实训,实训项目从简单的基本制作工艺着手,逐步过渡到西餐经典菜肴制作。

第二,实训项目中菜品的选择体现了特色性,突出了技能性。本教材紧跟西餐行业快速发展的步伐,除了传统的西餐菜肴外,还引入西餐最新菜品制作技术,使学生接触最新的专业知识,提升学生的专业洞察力。与此同时,结合现代人的膳食习惯,添加了西餐素食菜品制作技术,针对西餐行业发展的动向,增加了西式宴会小食制作技术等实训内容。

第三,体现了双语特色。本书部分理论章节及菜品制作环节采用了中英文双语穿插介绍的编写,体现了西餐烹饪的专业特色。

本书由余松筠任主编,丁辉、李俊(武汉市第一商业学校)任副主编,在编写过程中,参阅了已出版的相关教材,并得到了武汉商学院、华中科技大学出版社的大力支持,在此表示衷心的感谢。受作者水平所限,本书难免存在缺点或不当之处,恳请各位专家、教师和读者予以指正。本书由武汉商学院资助出版。

余松筠

Contents 目 录

第一章　西餐概述 ·· 1
　第一节　西方餐饮文化 ··· 1
　第二节　西餐发展概况 ··· 4
　第三节　西餐的主要菜式 ·· 7

第二章　西餐特色原料 ··· 11
　第一节　肉类原料 ··· 11
　第二节　肉制品 ·· 13
　第三节　家禽类 ·· 14
　第四节　水产类 ·· 16
　第五节　蔬果类 ·· 21
　第六节　奶类 ··· 30
　第七节　西餐调味品 ·· 39
　第八节　西餐香料 ··· 42
　第九节　意大利面食 ·· 47

第三章　西餐用具 ··· 57
　第一节　厨房用具 ··· 57
　第二节　西餐餐具 ··· 64

第四章　西式早餐 ··· 66
　第一节　西式早餐的分类及其特点 ·· 66
　第二节　西式早餐制作工艺 ··· 68

第五章　西餐少司的制作工艺 ·· 82
　第一节　少司的概念与作用 ··· 82
　第二节　少司的组成 ·· 83

1

第三节　少司的基本分类 ·· 85
　　第四节　少司制作实例 ·· 86

第六章　西式沙拉的制作工艺 ··· 96
　　第一节　沙拉的分类及其特点 ··· 96
　　第二节　沙拉的制作工艺 ·· 99

第七章　西式汤菜的制作工艺 ··· 106
　　第一节　基础汤 ·· 106
　　第二节　西餐汤的种类 ·· 110
　　第三节　汤的制作实例 ·· 112

第八章　西餐热菜的制作工艺 ··· 121
　　第一节　以油传热的烹调方法 ·· 121
　　第二节　用水传热的烹调方法 ·· 132
　　第三节　用空气传热的烹调方法 ····································· 139

第九章　西餐素食菜品的制作工艺 ··· 146
　　第一节　素食主义概述 ·· 146
　　第二节　西餐素食菜品制作 ··· 147

第十章　西式甜点的制作工艺 ··· 163
　　第一节　蛋糕的制作工艺 ··· 163
　　第二节　派类制作工艺 ·· 175
　　第三节　泡芙类制作工艺 ··· 179
　　第四节　冷冻类甜品的制作工艺 ····································· 180
　　第五节　水果甜品的制作工艺 ·· 185

第十一章　西式宴会小食 ·· 193

主要参考文献 ··· 202

彩图 ·· 203

第一章 西餐概述

第一节 西方餐饮文化

一、西餐的概念

西餐,是指西方国家的餐食,由其特定的地理位置所决定。我们通常所说的西餐是相对于中餐来说的,是对欧美各国菜点的统称,它不仅包括西欧国家的饮食菜肴、东欧各国的饮食菜肴,还包括美洲、大洋洲、中东、中亚、南亚次大陆以及非洲等地区的饮食。从广义上说,西餐是来自西方的菜点;从狭义上说,西餐由拉丁语系国家的菜肴组成。

据资料记载,西餐的起源有几个版本。在公元前5世纪,古希腊的西西里岛就开始讲究烹调方法,出现了较为原始的烹饪文化。在中世纪以后,法国人将西餐发挥至极致,他们在西餐烹饪和服务上进行考究,加之当时的几朝国王对西餐的重视,使得法式西餐套上了王宫的雍容华贵的光环,以至于后来西餐始终重视就餐礼仪、就餐环境及菜品创新,这也正是西餐的特色之处。

西餐作为欧美文化的一部分,具备以下几个特征。

(1) 西餐的就餐方式以刀、叉、匙为主要进食工具。西餐的餐具以刀、叉、匙为主,但由于食物的不同,又演变出了对应不同菜品的餐具,如沙拉刀叉、鱼刀叉、甜品刀叉、主餐刀叉等。

(2) 西餐的烹饪方法和菜点风味充分体现欧美特色。和中餐相比,西餐有其独特的烹饪方法,如焗是西餐中特有的烹调方法,其菜点口感、风味充分体现了欧美国家的特色。

(3) 西餐的服务方式、就餐习俗和情调充分反映欧美文化。西餐的服务方式指西餐用餐时提供给用餐者的侍应招待方式,服务方式视不同类型、特色、场合、消费而定。

吃西餐时讲究"6M",即 menu(菜谱)、music(音乐)、mood(气氛)、meeting(会面)、

manner(礼节)、meal(食品)。

菜谱被视为餐厅的门面,点菜是吃西餐的一个必不可少的程序,是一种优雅生活方式的表现。

豪华高级的西餐厅,通常会有乐队演奏一些柔和的乐曲,一般的西餐厅也会播放一些美妙典雅的乐曲。

吃西餐讲究环境雅致,气氛和谐。一般要有音乐相伴,桌台要保持整洁干净,餐具要洁净。如果是吃晚餐,灯光要暗淡,桌上要有红色蜡烛,从而营造一种浪漫、迷人、淡雅的气氛。

吃西餐主要是为联络感情,最好不要在西餐桌上谈生意。因此在西餐厅内,氛围一般比较温馨。

吃西餐讲究礼节,应遵循西方的习俗,不可有唐突之举,特别是在手拿刀叉时,若手舞足蹈,就会失态。刀叉的拿法一定要正确:应是右手持刀,左手拿叉。用刀将食物切成小块,然后用叉送入口内。一般来讲,欧洲人使用刀叉时不换手,一直用左手持叉将食物送入口内。美国人则是切好后,把刀放下,右手持叉将食物送入口中。

西餐宴会中主人都会安排男女相邻而坐,讲究女士优先。

二、西餐与中餐的区别

西餐进入中国后,经历了一百余年的发展历史。其间,西餐在中国饮食文化的影响下,逐渐地形成了一些共性。首先,两种饮食都是以补充人体膳食、改善营养结构为主要目的;其次,中餐与西餐不仅是改善人体机能的填充物,还在饮食文化上具有极高的一致性,它们分别承载了种族文化与国家文化的精髓,对于推进中西方文化建设与中西方文化交流具有积极的作用;再次,中餐与西餐在追求营养价值、精美的制作工艺以及改进和创新的方式上也有着极高的共同点。

由于受历史、地理、民族等多种因素的影响,中西餐也存在一些不同点,这主要体现在以下几个方面。

(一)饮食观念

中国饮食文化注重营养的搭配与手工艺的精湛,在制作中采用了煎、炸、烹、炒的方法,注重菜品的色香味俱全,体现了提高营养、全面搭配的饮食观念;而西餐注重热量的汲取,菜品制作比较简单,色泽、工艺没有中餐考究,西餐在注重简约的同时,更加注重饮食的健康与互补。

(二)饮食模式

西餐以富含蛋白质、脂肪、热量的肉食为主,含碳水化合物、纤维素成分的食物偏少,而中餐主要由谷物提供热量。在烹调方式上,西餐采用机器操作进行大规模化生产,要有营养、方便、快捷;中餐则注重细火慢温,把菜肴做得精细,以调和五味为根本,以色彩艺术为精华,要求色、香、味、形俱佳。

(三) 饮食结构

西餐以肉类食品为主,蔬菜和米面通常用以佐餐,且菜肴多以大块和整体出现,进餐工具为刀叉,同时促成了分餐的良好的进餐方式。分餐制既有利于个人口味的选择,增加了进餐的自由度,又可避免无谓的浪费,降低病菌交叉感染的机会。

(四) 选料的区别

中餐的选料非常广泛;西餐常用的原料有牛肉、羊肉、猪肉和禽类、乳蛋类等,很少使用动物内脏。

(五) 原料加工的区别

中餐厨师非常讲究刀工,常把原料加工成细小的丝、丁、片、末等;而西餐厨刀的种类多,用法讲究,但很少把原料加工成细小的形状,大都是体积较大的排、卷、块等,造型别致。

(六) 烹调方法的区别

中餐烹调时一般使用圆底锅、明火灶,中餐采用炒的烹调方法较多;而西餐使用平底锅、暗火灶、扒板、面火炉等设备,烹调方法主要是煎、烤、焖、烩、铁扒等。

(七) 口味的区别

中餐菜肴大都有明显的咸味,并富于变化,多数菜肴都是完全成熟后再食用。西餐菜肴很少有明显的咸味,口味变化较少,但追求菜肴鲜嫩的效果,比如牛排、羊排等菜肴要根据客人的需求来确定其成熟度。西餐非常讲究少司的制作,很多菜都配有少司,用来增加菜肴的口味。

(八) 主食的区别

中餐有明确的主、副食概念;而西餐并无明确的主、副食概念,面包及其他面食、米饭经常作为配菜放在盘子旁边,用量较少。

(九) 上菜顺序的区别

中餐上菜顺序:先上冷菜、饮料及酒,后上热菜,然后上主食,最后上甜食和水果。
西餐上菜程序:开胃菜→汤→副菜→主菜→甜品→咖啡或茶。

第二节 西餐发展概况

一、西餐在西方的发展概况

在公元前5世纪,古希腊的西西里岛就开始讲究烹调方法,出现了较为原始的烹饪文化,当时不仅已经有了煎、炸、烤、焖、煮、炙、熏等多种烹调方法,还发明了水果蛋糕和杏仁蛋糕塔,同时技术高超的厨师非常受社会的尊敬。当时的古罗马文化还比较落后,后来随着其疆域的不断扩大,逐渐吸收了当时先进的古希腊文化,餐饮文化开始被重视起来,并很快发展到一个新水平。后来古罗马宫廷的膳房已形成庞大的规模,并分有面包、水果、菜肴、葡萄酒4个专业,厨师总管的身份与贵族大臣相同。这时的烹调方法也日臻完善,并发明了数十种少司的制作方法。这些餐饮文化后来极大地影响了大半个欧洲,被誉为"欧洲大陆烹饪之始祖"。

西罗马帝国灭亡后,居住在北欧的日耳曼族各部落向南迁徙,在其废墟上逐渐建立起一系列封建割据国家,并与当地居民逐渐融合。日耳曼人以肉食为主,其饮食习惯影响了西方各国,在前后大约一千年的时间内,欧洲大部分地区的餐饮文化发展得比较缓慢。直到15世纪中叶欧洲文艺复兴时期,餐饮文化以意大利为中心发展起来,在当时贵族举办的宴会上不断出现各种名菜、甜点。但餐具的使用普及较晚,在公元6世纪,东罗马帝国发明了两齿的叉子,但得不到普及,意大利的大主教认为"只有上帝创造的人类手指才配接触上帝的赐物"。直到18世纪,随着人们观念的转变,餐叉、餐刀的使用才得到普及,同时餐叉也由2齿变为4齿,餐刀也由尖头变为圆头。

瓷器的使用也较晚,到1710年德国才出现欧洲最早的瓷窑,但发展很快,到18至19世纪,在近代自然科学和工业革命的影响下,餐饮文化也发展到一个崭新阶段,发明了很多先进的炊具和餐具,社会上也涌现出大量的饭店和餐厅,形成了高度发展的餐饮文化。

20世纪是西餐发展的鼎盛时期,一方面,原来的宫廷大菜已逐渐在民间普及;另一方面,西餐也朝着个性化、多样化发展,品种更加丰富。与此同时,由于工业化的迅速发展,食品工业也随之产生,并逐渐形成完整的体系。快餐是西方餐饮发展史上的一次变革,它是烹饪的工业化。50年代,由于战后经济的迅速发展,女性就业人数的增加,人们生活方式的改变,快餐首先在美国产生。由于快餐适应快节奏的生活方式,加之现代化的经营理念和标准化的生产方法使快餐业有了极为广阔的发展空间,因此到60年代末、

70年代初美国的快餐业发展速度达到最高峰。受美国的影响,西方其他国家的快餐业也逐渐发展起来。

现代的西餐是西式正餐和西式快餐并存的模式,在知识和技术上都已经达到了成熟的境界。一方面,新的烹饪设备、新的烹调方法和新的原料被应用在烹调过程中;另一方面,随着营养学的发展,现代西餐变得更为科学合理。

西餐是随着西方强势文化的传播而传播全球的。近现代是西方科学技术蓬勃发展的时期,西餐作为强势文化在全球极力推广。西餐的发展有了广阔的市场空间,无论是烹调的原料、烹调的能源、烹调的器械及设备,还是烹调的管理、烹调的标准等,西餐都有了长足的发展。

二、西餐在我国的传播与发展

中国菜品在古代几千年传承的基础上,一直保持着传统的制作方法,体现了各地区的饮食菜品特色。近代以来,中国也吸收了西方的菜式和菜品的制作方法。1949年以后,中国的菜品为了适应和满足外国客人的口味,在制作技艺上不断吸收外国菜品技术。进入21世纪以来,中西餐烹饪技艺已进一步融合发展,从原料的引用、调味品的引入,到烹饪技法的借鉴,中餐菜品不断增加新的内容和新的制作思路。此外,随着科学技术的进步,炊事用具得到不断改进并更新,不锈钢制品得到普及,厨房变得光洁明亮。电动粉碎机、搅拌器、高压锅、切片机,还有自动控温的油炸炉等,使得蔬菜的切片、切丝加工等都可用机器来完成,西餐炊具、餐具也影响并改变着中式烹饪的传统方式。

西餐在中国的传播和发展经历了以下几个阶段。

(一) 17世纪中叶以前一些西方食品的传入

西餐在我国有着悠久的历史,它是伴随着我国与世界各国人民的交往而传入的,但西餐到底何时传入我国,至今还未有定论。据史料记载,早在汉代,波斯古国和西亚各地的灿烂文化通过"丝绸之路"传到中国,其中就包括膳食。在元代,意大利旅行家马可·波罗在我国居住数十年,为两国经济文化交流做出了重大贡献。他曾把中国面条带到意大利,经意大利人民改制,创造出了举世闻名的意大利面条。与此同时,马可·波罗也给成吉思汗的子孙带来了意大利的佳肴美馔。但是在漫长的封建社会中,中西方的交往是十分有限的,当时在食品方面,只限于一些物产的相互交流,如西方的芹菜、胡萝卜、葡萄酒等陆续传入我国。

(二) 17世纪中叶至19世纪初西方饮食习俗的传入

17世纪中叶,西欧开始出现资本主义,一些商人为了寻找商品市场,陆续来到我国广东、福建等沿海地区经商,一些政府官员和传教士也先后到我国部分城镇进行传教等活动。这些人一般在我国居住的时间较长，由于生活上的习惯，他们自带本国食品和本国厨师，也有的雇佣中国人为他们服务，这样西方国家的生活方式对我国就产生了较大的影响，但当时这样简单的西餐也只能在来华的外国人的家庭餐桌上出现。随着涌入的商人、传教士等外国人的增多，中国宫廷、王府官吏开始与西方人交往频繁，他们逐步对西餐产生兴趣，有时也吃起西餐。但当时，我国的西餐业还没有形成。

（三）19世纪中叶至20世纪三四十年代西餐馆的发展及其影响

鸦片战争后，来华的西方人越来越多，从而把西方饮食的烹饪技艺带到了中国。

清代光绪年间，在外国人较多的上海、北京、广州、天津等地，出现了由中国人经营的西餐厅（当时称"番菜馆"），以及咖啡厅、面包房等，从此我国有了西餐行业。据清末史料记载，最早的番菜馆是上海福州路的"一品香"，在北京最早出现的是"醉琼林"、"裕珍园"等。

20世纪二三十年代是西餐在中国传播和发展的最快时期，在上海出现了几家大型的西式饭店，如礼查饭店（现浦江饭店）、汇中饭店（现和平饭店南楼）、大华饭店等。进入30年代，又相继出现了国际饭店、上海大厦等。这些饭店除接待住宿以外，都以经营西餐为主。

（四）1949年至十一届三中全会前西餐的延续

西餐在我国几经盛衰。至1949年前夕，由于连年战乱，西餐业已濒临绝境，从业人员所剩无几。1949年后，历史赋予西餐业新的生命。随着我国国际地位的提高，世界各国与我国的友好往来日益频繁，我国又陆续建起了一些经营西餐的餐厅、饭店，如北京的北京饭店、和平饭店、友谊宾馆、新侨饭店等。20世纪50年代和60年代我国的西餐以俄式菜发展较快。

20世纪70年代初，我国餐饮市场经历了严重的萎缩阶段，西式餐厅衰退得更厉害，有的关闭，有的转行，勉强维持经营的也很难保持特色。但西餐在中国始终没有消失过。

（五）十一届三中全会至今西餐的空前繁荣

西餐业在中国的振兴应该是20世纪80年代。党的十一届三中全会确定了中国实行改革开放，国民经济迅速增长，人民生活不断改善，外国来访人员和旅游外宾的人数急剧增加，使西餐业在中国迎来了第二个大发展时期。随着国际旅游饭店的大量兴建，肯德基、麦当劳等西式快餐的涌入，西餐已进入中国的各个城市，其中不少饭店系中外合资或外资企业，不但聘用了不少外国厨师，而且引进了不少的新设备和新技术。与此同时，原来的一些老饭店也在不断改进，陆续派厨师去国外学习。如今，西餐越来越为人们所了解，它以科学的营养搭配、干净的就餐环境、奇特的饮食风味、浓烈的异国情调吸引着我国人民，丰富了人们的日常饮食生活。同时，在全球文化的深度传播交流下，我国的青年一代希望通过尝试异国饮食的方式来融入国际化的文化氛围,西餐的出现为此提供了一个较容易被接受的平台,这在某种程度上也促进了西餐在我国的发展。

第三节 西餐的主要菜式

西餐在长期的发展过程中,由于不同国家地区的自然条件、历史原因、社会制度、饮食文化、口味等不同形成了很多不同菜式,其中影响较大的有法国菜、意大利菜、美国菜、英国菜、德国菜、俄罗斯菜等。

一、法国菜

法国菜是世界上著名的菜系之一,已为众所公认。它的口感之细腻,酱料之美味,餐具摆设之华美,简直可称之为一种艺术。

法国菜的文化源远流长,相传在16世纪意大利女子Catherine嫁给法国国王亨利二世以后,把意大利文艺复兴时期盛行的牛肝脏、黑菌(黑松露)、嫩牛排、奶酪等的烹饪方法带到法国,路易十四还曾发起烹饪比赛。17世纪后,法国菜不断地精益求精,并将以往的古典菜肴推向新菜烹调法,并相互运用,调制的方式讲究风味、天然性、技巧性、装饰和颜色的配合。

法国菜有很多特点,主要体现在以下几个方面。

(1) 选料广泛、讲究。法国菜的突出特点是选料广泛。法国菜常选用稀有的名贵原料,如蜗牛、鹅肝、黑松露等。在法国,鱼子酱、鹅肝、松露被誉为"西方三大珍味"。

(2) 重视调味,调味品种类多样。法国盛产酒类,烹调过程中喜欢用酒调味,酒和菜品的搭配都有严格的规定。

(3) 使用新鲜的季节性食材。法国美食的特色在于使用新鲜的季节性材料,加之厨师个人的独特的调理,能完成独一无二的艺术极品佳肴,无论是在视觉、嗅觉、味觉,还是在触感上,都是无与伦比的境界。

(4) 讲究菜肴的鲜嫩度。法国菜比较讲究吃半熟食物或生食,如牛肉、羊肉通常烹调至六七分熟即可,海鲜烹调时须熟度适当,不可过熟,而牡蛎大都生吃。

(5) 服务形式多样、用餐环境讲究。法式服务形式多样,例如:在边桌上分食,使食物在烹饪过程中起火焰,这些服务方式能更好地从视觉上使客人满意,让客人更好地享受食物带来的乐趣。法餐在就餐环境和气氛上,要求更精致化,如法国菜在享用时非常注重餐具(刀、叉、盘、酒杯)的使用,因为这些均可衬托出法国菜高贵的气质。

典型的法国菜有：鹅肝酱、法式焗蜗牛、洋葱汤、马赛鱼羹等。

二、意大利菜

意大利优越的地理条件使其农业和食品工业都很发达，其中以面条、奶酪、色拉米肉肠著称于世，意大利菜在世界上享有很高的声誉。

意大利菜源自古罗马帝国宫廷，有着浓郁的文艺复兴时代佛罗伦萨的膳食情韵。自公元前753年罗马城兴建以来，古罗马帝国在吸取了古希腊文明精华的基础上，发展了先进的古罗马文明，以佛罗伦萨城为首的王公贵族们，纷纷以研究开发烹调技艺及是否拥有厨艺精湛的厨师来展现自己的实力与权力，并以此为荣耀。

意大利菜对欧美国家的餐饮产生了深远影响，并发展出包括法国餐、美国餐在内的多种派系，故有"西餐之母"的美称。

意大利烹饪以世界精美菜肴著称，具有自己的风格特色。

（1）意大利菜注重原汁原味，讲究火候的运用。意大利菜注重原料的本质、本色，成品力求保持原汁原味。在烹煮过程中非常喜欢用蒜、葱、西红柿酱、干酪，讲究制作少司。烹调方法以炒、煎、烤、红烩、红焖等居多。通常将主要材料或裹或腌，或煎或烤，再与配料一起烹煮，从而使菜肴的口味异常出色，缔造出层次分明的多重口感。

（2）巧妙利用食材的自然风味。橄榄油、黑橄榄、干白酪、香料、西红柿与马沙拉（Marsala）酒，这六种食材是意大利菜烹制的灵魂，也代表了意大利当地所盛产与充分利用的食用原料，因此意大利菜被称为"地道与传统"。最常用的蔬菜有西红柿、白菜、胡萝卜、龙须菜、莴苣、土豆等。意大利人对肉类的制作及加工非常讲究，如风干牛肉（dry beef）、风干火腿、意大利腊肠、波伦亚香肠、腊腿等。

（3）以米面做菜，花样繁多，口味丰富。意大利人擅长做面类、饭类制品，几乎每餐必做，而且品种多样，风味各异。著名的有意大利面、比萨饼等。意大利面具有不同的形状和颜色，斜状的是为了让酱汁进入面管中，而条纹状的是为了让酱汁留在面条表层上，颜色则代表了面条添加了相应的营养素。

典型的意大利菜有：意大利菜汤、米兰式猪排、奶酪焗通心粉、比萨饼、红花米饭等。

三、美国菜

自从哥伦布1492年发现美洲大陆后，欧洲的一些国家就开始不断向北美移民，在此开拓殖民地。至1733年，英国人已在北美建立了13个殖民地。在开发当地经济的同时，他们也把原居住地的生活习惯、烹调技艺等带到了美国，因此美国菜可称得上是东西交汇、南北合流。

但因为其中大部分居民都是英国人，且到了17世纪和18世纪后期，美国受英国统治,所以英式文化在这里占统治地位。美国菜主要是在英国菜的基础上发展而来的,另外又融合了法、意、德等国家的烹饪精华,兼收并蓄,形成了自己的独特风格。其主要特点体现在以下几个方面。

(1) 喜欢用水果做菜。美国盛产水果,美国菜的沙拉中水果用得较多,如用香蕉、苹果、梨、橘子等做沙拉最为普遍。另外,在热菜中也常使用水果,如菠萝焗火腿、苹果烤火鸡、炸香蕉等。

(2) 口味清淡、生鲜。由英国菜派生出来的美国菜发展至今,在口味及用料上已经发生了不少变化。传统的咸鲜甜口味已趋向清淡、生鲜。在用料上,黄油改用植物黄油,奶油改用完全脱脂奶油,奶酪改用液态奶,生菜沙拉不用马乃司少司,浓汤改清汤,肉类则多用低脂及低胆固醇的水牛肉等。另外,在美国,素食和生食比较盛行。

(3) 快餐食品发达。因为美国传统、保守的思想较少而经济比较发达,所以快餐业首先在美国应运而生,并很快影响到世界各地的餐饮业。

美国菜代表菜有:华尔道夫沙拉、烤火鸡配苹果、菠萝火腿扒、美式牛扒、苹果派等。

四、英国菜

英国的农业不发达,粮食每年都要进口,也没有法国人那样崇尚美食,因此英国菜相对来说比较简单,但英式早餐却很丰盛,素有"big breakfast"即丰盛早餐的美称,受到西方各国的普遍欢迎。英国菜的主要特点体现在以下几个方面。

(1) 选料局限。英国菜选料比较简单,受地理自然条件所限,渔场不太好,因此英国人不讲究吃海鲜,比较偏爱牛肉、羊肉、禽类等。

(2) 口味清淡、原汁原味。简单而有效地使用优质原料,并尽可能保持其原有的质地和风味是英国菜的重要特色。英国菜的烹调对原料的取舍不多,一般用单一的原料制作,要求厨师不加配料,保持菜式的原汁原味。

英国菜有"家庭美肴"之称,英国烹饪法根植于家常菜肴,因此只有原料是家生、家养、家制时,菜肴才能达到满意的效果。

(3) 烹调简单、富有特色。英国菜的烹调相对来说比较简单,配菜也比较简单,香草与酒的使用较少,常用的烹调方法有煮、烩、烤、煎、蒸等。

(4) 早餐丰盛。英式早餐非常丰盛,一般有各种蛋品、麦片粥、咸肉、火腿、香肠、黄油、果酱、面包、牛奶、果汁、咖啡等。

英国菜代表菜有:土豆烩羊肉、烤鹅填栗子馅、牛尾浓汤、皇家奶油鸡等。

五、德国菜

德国人喜爱运动,食量较大,以肉食为主,德式菜肴口味较重,材料上则偏好猪肉、牛肉、香料、鱼类、家禽及蔬菜等,使用大量芥末、白酒、牛油等,而在烹调上较常使用煮、炖或烩的方式。

（1）肉制品丰富。德国的肉制品种类繁多，德国菜中有不少是用肉制品制作的菜肴，仅香肠一类就有上百种，法兰克福肠早已驰名世界。

（2）喜欢食用生鲜。德国人有吃生牛肉的习惯，著名的鞑靼牛扒就是将嫩牛肉剁碎，拌以生葱头末、酸黄瓜末和生蛋黄食用。

（3）口味以酸咸为主。德国菜中的酸菜使用非常普遍，经常用来做配菜，口味酸咸，浓而不腻。

（4）用啤酒制作菜肴。德国盛产啤酒，啤酒的消费量也居世界之首，因此菜肴也常用啤酒来调味。

德国典型的菜式有：酸菜焖法兰克福肠、汉堡肉扒、鞑靼牛扒等。

六、俄罗斯菜

作为一个地跨欧亚大陆的世界上领土面积最大的国家，俄罗斯虽然在亚洲的领土非常辽阔，但由于其绝大部分居民居住在欧洲地区，因而其饮食文化更多地受到了欧洲大陆的影响，呈现出欧洲大陆饮食文化的基本特征。特殊的地理环境、人文环境以及独特的历史发展进程，造就了独具特色的俄罗斯饮食文化。

（1）传统菜油脂较大。由于俄罗斯气候寒冷，人们需要补充较多的热量，俄罗斯菜一般用油比较多，多数汤菜上都有浮油。随着社会的进步，人们的生活方式也在发生改变。到了20世纪六七十年代后，俄罗斯菜也逐渐趋于清淡。

（2）口味浓厚。俄罗斯菜口味浓厚，而且酸、甜、咸、辣俱全，人们喜欢吃大蒜、葱头。

（3）讲究冷小吃。俄罗斯小吃是指各种冷菜，其特点是生鲜、味酸咸，如鱼子酱、酸黄瓜、冷酸鱼等。

典型的俄罗斯菜有：鱼子酱、红菜汤、黄油鸡卷、莫斯科烤鱼等。

第二章　西餐特色原料

第一节　肉类原料

肉类是西餐中主要的原料。肉的品质取决于动物的种类及饲养方式。肉类原料在西餐烹调应用中的菜式较多、风味各不相同，菜品风味受肉的部位、加热方式、烹调温度与烹调时间等多种因素的相互影响。

一、牛肉(beef)

牛肉的品种（色泽、外观、脂肪含量、肉的组织）取决于牛的年龄，畜龄越大，肌肉质地越粗糙，嫩度也有所降低，牛肉中大理石纹脂肪含量越高，分布越细致均匀，牛肉的风味、含汁量与口感就会越好。

西餐中牛肉的使用很讲究，一般把牛肉分为五级，根据不同的肉质适当选用。

特级肉是指牛的里脊(tenderloin)，因为这个部位很少活动，所以肉质纤维细软，是牛肉中最嫩的部分，常用来做菲力牛排及铁板烧。菲力牛排对操作要求比较高，一般煎至3~5成熟，以此保持肉的鲜嫩多汁。

一级肉是牛的脊背部分，包括外脊和上脑两个部位。外脊(sirloin)也称西冷，是牛背部的最长肌，肉质为红色，容易有脂肪沉积，呈大理石花纹状，我们常吃的西冷牛排就是用这部分的肉。因为外脊含有脂肪，所以煎、烤起来味道、口感更佳。上脑肉质细嫩，容易形成大理石花纹，脂肪交杂均匀，适合涮、煎、烤等。

二级肉是指牛后腿的上半部分。

三级肉包括前腿、胸口和肋条肉。前腿肉纤维粗糙，肉质老硬，一般用于绞馅，做各种肉饼。胸口和肋条肉质虽老，但肥瘦相间，味道鲜美，用来做焖牛肉、煮牛肉比较合适。

四级肉包括脖颈、肚脯和腱子肉。这部分肉筋皮较多，肉质粗老，适宜煮汤。牛颈肉肥瘦兼有、肉质干实、肉纹较乱，适宜制馅或煨汤。腱子肉分为前腱和后腱，熟后有胶质感,适合红烧或卤。

牛的尾巴筋皮多,有肥有瘦,可用来做汤、烩牛尾等。

小牛肉又称牛仔肉、牛犊肉,是指生长期在3～10个月时宰杀获得的牛肉,其中饲养3～5月龄的又称为乳牛肉或白牛肉,英文称为white veal。饲养5～10月龄的称为小牛肉或牛仔肉。若小牛生长期不足3个月,其肉质中水分太多,不宜食用。3个月以后,小牛肉肉质渐纤细,味道鲜美,特别是3～5月龄的乳牛,由于此时尚未断奶,其肉质更是细嫩、柔软,富含乳香味。一般小牛的生长期过了12个月,则肉色变红,纤维逐渐变粗,此时就不能再叫作小牛肉。小牛肉肉质细嫩、柔软,脂肪少,味道清淡,是一种高蛋白、低脂肪的优质原料,在西餐烹调中应用广泛,尤其在意大利菜、法国菜中更为突出。

小牛除了部分内脏外,其余大部分部位都可以作为烹调原料,特别是小牛喉管两边的膵脏,又称牛核,被视为西餐烹调中的名贵原料。

二、羊肉(mutton)

在西餐烹调中,羊肉的应用仅次于牛肉。羊在西餐烹调上又有羔羊(lamb)和成羊(mutton)之分,羔羊是指生长期在3个月至1年的羊,其中没有食过草的羔羊又被称为乳羊(milk fed lamb),成羊是指生长期在1年以上的羊。西餐烹调中主要以使用羔羊肉为主。

羊的种类很多,其品种类型主要有绵羊、山羊和肉用羊等,其中肉用羊的羊肉品质最佳,肉用羊大都是用绵羊培育而成,其体型大,生长发育快,产肉性能高,肉质细嫩,肌间脂肪多,切面呈大理石花纹状,其肉用价值高于其他品种。澳大利亚、新西兰等国是全世界主要的肉用羊生产国,目前我国的市场供应主要以绵羊肉为主,山羊肉因其膻味较大,故相对较少。

三、猪肉(pork)

猪肉也是西餐烹调中最常用的原料,尤其是德国菜对猪肉更是偏爱,其他欧美国家也有不少菜肴是用猪肉制作的。

猪在西餐烹调上有成年猪(pig)和乳猪(sucking pig)之分。乳猪是指尚未断奶的小猪。乳猪肉嫩色浅,水分充足,是西餐烹调中的高档原料。成年猪一般以饲养1～2年为最佳,其肉色淡红,肉质鲜嫩、味美。在西餐烹调过程中,从食品安全与卫生角度考虑,猪肉必须全熟。

第二节 肉制品

肉制品(meat products),是指用畜禽肉为主要原料,运用物理或化学的方法,配以适当辅料和添加剂,对原料肉进行工艺处理后得到的产品。西方国家食品工业比较发达,畜肉制品的种类丰富,如香肠、火腿、培根、熏肉等,这些食品食用方便,在西餐中被广泛应用,常用来制作开胃菜、沙拉等。

一、火腿(ham)

火腿是一种常见的肉制品。火腿一般是用猪的前后腿部肉经腌制、洗晒、整形、陈放发酵等工艺加工而成的腌制品。西式火腿可分为两种类型:无骨火腿和带骨火腿。

火腿在西餐烹调中被广泛使用,可以用作主料、辅料和制作冷盘。

火腿的保藏主要是避免油脂酸败、回潮发雾、虫蛀等,应将火腿放在阴凉、干燥、通风、清洁处,避免高温和阳光照射,隔氧。

二、培根(bacon)

培根由英语"bacon"音译而来,培根是西式肉制品三大主要品种(火腿、香肠)之一,其风味除带有适度的咸味之外,还具有浓郁的烟熏香味。培根外皮油润呈金黄色,皮质坚硬,瘦肉呈深棕色,质地干硬,切开后肉色鲜艳。

根据其制作原料和加工方法的不同主要分为以下几种。

(一) 五花培根

五花培根也称美式培根(American bacon),是将猪五花肉切成薄片,用盐、亚硝酸钠或硝酸钠、香料等腌渍、风干、熏制而成。

(二) 外脊培根

外脊培根也称加拿大式培根(Canadian bacon),是用纯瘦的猪外脊肉经腌渍、风干、熏制而成,口味近似于火腿。

(三) 爱尔兰式培根

爱尔兰式培根(Irish bacon)是用带肥膘的猪外脊肉经腌渍、风干加工制成的,这种培根不用进行烟熏处理,肉质鲜嫩。

(四)意大利培根

意大利培根(Italian bacon)意大利文为"Pancetta",是将猪腹部肥瘦相间的肉,用盐和特殊的调味汁等腌渍后,将其卷成圆桶状,再经风干处理,切成圆片制成的。意大利培根也不用进行烟熏处理。

三、香肠(sausage)

香肠的种类很多,仅西方国家就有上千种,主要有冷切肠系列、早餐香肠系列、色拉米肠系列、小泥肠系列,风干肠、烟熏香肠及火腿肠系列等。其中生产香肠较多的国家有德国和意大利等。

制作香肠的原料主要有:猪肉、牛肉、羊肉、火鸡肉、鸡肉和兔肉等。其中以猪肉的使用最为普遍。一般的加工过程是先将肉绞碎,加上各种不同的辅料和调味料,然后灌入肠衣,再经过腌渍或烟熏、风干等制成。

世界上比较著名的香肠品种有:德式小泥肠(Bratwurst)、米兰色拉米香肠(Milan salami)、维也纳牛肉香肠(Viennese sausage)、法国香草色拉米香肠(French herb salami)等。在西餐烹调中可用香肠做沙拉、三明治、开胃小吃及煮制菜肴,也可将其作为热菜的辅料。

第三节 家 禽 类

家禽是西餐中的重要原料,主要品种有鸡、鸭、鹅、鸽子等。西餐中经常将禽肉分为白色肉类、深色肉类等,白色肉类肉质呈白色,含脂肪和结缔组织较少,烹调时间短,如鸡肉;深色肉类肉质呈褐色,脂肪和结缔组织较多,烹调时间长,如鸭肉、鹅肉、鸽子肉。

一、鸡(chicken)

鸡的种类较多,主要有以下种类。

(一)雏鸡

雏鸡是指生长期在1个月左右,体重为250~500 g的小鸡。雏鸡肉虽少,但肉质鲜嫩，适宜整只烧烤、铁扒等。

（二）春鸡

春鸡又名童子鸡，是指生长期在两个半月左右，体重为500～1250 g的鸡。春鸡肉质鲜嫩，口味鲜美，适宜烧烤、铁扒、煎、炸等。

（三）阉鸡

阉鸡又称肉鸡，是指生长期在3～5个月，用专门的饲料喂养的，体重为1500～2500 g的公鸡。阉鸡肉质鲜嫩，油脂丰满，水分充足，但由于其生长期较短，香味不足，适宜煎、炸、烩、焖等。

二、鸭（duck）

鸭从其主要用途看，可将其分为羽绒型、蛋用型、肉用型鸭等品种，西餐烹调中主要使用肉用型鸭作为烹调原料。

肉用型鸭饲养期一般为40～50天，体重可达2500～3500 g。肉用型鸭胸部肥厚，肉质鲜嫩。鸭在西餐中的使用也很普遍，常用的烹调方法主要有烤、烩、焖等。鸭肝可以制作各种"肝批"。

三、鹅（goose）

鹅在世界范围内的饲养很普遍，从其主要用途看，鹅的品种可分为羽绒型、蛋用型、肉用型、肥肝用型等，与西餐烹调有关的主要是肉用型和肥肝用型鹅。

肥肝用型鹅主要是应用其肥大的鹅肝。这类鹅经填饲后的肥肝重达600 g以上，优质的则可达1000 g左右，其著名的品种主要有法国的朗德鹅、图卢兹鹅等，这类鹅也可用于产肉，但习惯上把它们作为肥肝专用型品种。肥鹅肝是西餐烹调中的上等原料，在法国菜中的应用最为突出，鹅肝酱、鹅肝冻等都是法国菜中的名菜。由于鹅肝中含有大量脂肪，因此在烹调时不要用急火，以免脂肪流失，使鹅肝质地变干。优质的鹅肝具备以下特点。

（1）颜色：呈乳白色或白色，其中的筋呈淡粉红色。

（2）硬度：肉质紧，用手指触压后不能恢复原来的形状。

（3）质感：肉质细嫩光滑，手触后有一种黏糊糊的感觉。反之，手感不光滑并发干，是质量较差的肥鹅肝。

肥鹅肝原则上应立即使用，不宜保存，如需要保存，应将肥鹅肝放进真空薄膜中，封口后放置于冰水中。

四、鸽子(pigeon)

鸽子又称家鸽,由岩鸽驯化而来,经长期选育,目前全球鸽子的品种已达1500多种,按其用途可分为信鸽、观赏鸽和肉鸽。西餐烹调中主要以肉鸽作为烹调原料。

肉鸽体型较大,一般雄鸽可重达500~1000 g,雌鸽也可达400~600 g。肉鸽肉色深红,肉质细嫩,味道鲜美。经专家测定,肉鸽一般在28天左右就能达到500 g左右,这时的鸽子是最有营养的,其含有17种以上的氨基酸,氨基酸总量高达53.9%,且含10多种微量元素及多种维生素。因此鸽肉是蛋白质含量高、脂肪含量低的理想原料。鸽子在西餐烹调中常被用于烧烤、煎、炸、红烩或红焖等,乳鸽一般适宜用于铁扒等。

五、火鸡(turkey)

火鸡原产于北美,一般在西餐中作为烹调原料使用的主要是肉用型火鸡,如美国的尼古拉火鸡。这些肉用火鸡的胸部肌肉发达,腿部肉质丰厚,生长快,出肉率高,脂肪含量低,胆固醇含量低,蛋白质含量高,味道鲜美,是西餐烹调中的高档原料,也是许多欧美国家在圣诞节、感恩节时餐桌上不可缺少的食品。

火鸡在体型上一般有小型火鸡、中型火鸡和重型火鸡之分。一般小型火鸡的体重为3~5 kg,中型火鸡的体重为6~9 kg,重型火鸡的体重为10~15 kg。一般体型较小、肉质细嫩的火鸡,适宜整只进行烧烤或瓤馅等。体型较大、肉质较老的火鸡,适宜烩、焖或去骨制作火鸡肉卷等。

第四节 水 产 类

水产类原料的特点是水分充足,味道鲜美,在烹调时水分损失较少,用各种海味烹制的菜肴肉质松软,易于消化,深受人们的欢迎。水产类食物在西餐中占有重要位置。从营养学的角度来看,水产类的营养价值是较为理想的。

一、鱼类

(一) 三文鱼(salmon)

三文鱼又名细鳞鱼,或红点鲑鱼,与大马哈鱼同属硬骨鱼纲鲑科,是鲑鱼的一种,主要产于美国、加拿大、挪威及英国的河口处等冷水域。

三文鱼是世界名贵鱼类之一。其鳞小刺少,肉色橙红,肉质细嫩鲜美,口感爽滑,既可直接生食,又能烹制菜肴,是深受人们喜爱的鱼类,同时由它制成的鱼肝油更是营养佳品。

选用时选取鱼鳃鲜红、鱼鳞银白的新鲜原料。

三文鱼是西方国家普遍喜欢食用的鱼种,因此在西餐中广泛使用,主要用于制作开胃菜中的冷盘、主菜(可烧烤、煎炒)等。

Salmon is a popular food, it is considered to be healthy due to the fish's high content of protein, omega-3 fatty acids and vitamin D. Salmon flesh is generally orange to red.

(二) 鳕鱼(gadus)

鳕鱼学名 gadus,俗称 cod,原产于从北欧至加拿大及美国东部的北大西洋寒冷水域。目前鳕鱼主要出产国是加拿大、冰岛、挪威、俄罗斯、日本。鳕鱼是全世界年捕捞量最大的鱼类之一,具有重要的食用价值和经济价值。

鳕鱼体色多样,从淡绿或淡灰到褐色或淡黑,也可为暗淡红色到鲜红色,体上有深色斑点。鳕鱼肉紧实、洁白、刺少,肉多味美。

鳕鱼肉味甘美、营养丰富,蛋白质含量高,所含脂肪的量和鲨鱼一样只有 0.5%,是三文鱼的 1/17。鳕鱼的肝脏含油量高,除了富含普通鱼油所有的 DHA、EPA 外,还含有人体所必需的维生素 A、D、E 和其他多种维生素。鳕鱼肝油中这些营养成分的比例,正是人体每日所需要量的最佳比例,因此,北欧人将它称为餐桌上的"营养师"。

鳕鱼可以用多种方式进行烹制,若蘸调味汁食用,味道则尤为鲜美。鳕鱼可被制成鱼肉罐头、鳕鱼干或腌熏鱼,鳕鱼子可新鲜食用,也可熏制或腌制。

挑选时应选择腹部有弹性、表皮呈淡褐色且有光泽的鳕雪。鳕鱼鲜度下降很快,挑选时应注意其新鲜度。

Gadus, which is commonly known as cod, is popular as food with a mild flavour and a dense, flaky, white flesh. Cod livers are processed to make cod liver oil, an important source of vitamin A, vitamin D, vitamin E, and omega-3 fatty acids (EPA and DHA).

In the United Kingdom, cod is one of the most common ingredients in fish and chips, along with haddock and plaice. Cod's soft liver can be tinned(canned) and eaten. Cod is mainly consumed in Portugal, Spain, Italy and Brazil.

(三) 金枪鱼(tuna)

金枪鱼又叫鲔鱼,香港称吞拿鱼,澳门以葡萄牙语旧译为亚冬鱼,大部分属于金枪鱼属。金枪鱼是大洋暖水性洄游鱼类,主要分布于低中纬度海区,在太平洋、大西洋、印度洋都有广泛的分布,我国东海、南海也有分布。金枪鱼的肉色为红色,这是因为金枪鱼的肌肉中含有了大量的肌红蛋白所致。

由于金枪鱼必须时常保持快速游动的状态才能维持身体的能量供给,加上其只在海域深处活动,因此金枪鱼的肉质柔嫩鲜美,且不被环境污染,是现代人不可多得的健康美食。金枪鱼的蛋白质含量高达20%,但脂肪含量很低,俗称海底鸡,营养价值高。鱼肉中的脂肪酸大多为不饱和脂肪酸,所含氨基酸齐全,其中包括人体所需的8种氨基酸,还含有维生素,以及丰富的铁、钾、钙、镁、碘等多种矿物质和微量元素。

金枪鱼可生食、熟食(铁扒、炸、煎、烧烤),也可制成油浸金枪鱼罐头。因为金枪鱼处在食物链的较顶端,体内的重金属含量和水银含量都要高于其他海鱼,所以不能过多食用。

鲜鱼选择鱼眼清亮、鱼皮鲜艳者。

When tuna is served as a steak, the meat of most species is known for its thickness and tough texture.

Tuna can be canned in edible oils, in brine, in water, and in various sauces. Tuna may be processed to be chunked or flaked. In the United States, 52% of canned tuna is used for sandwiches, 22% for salads and 15% for casseroles.

(四)沙丁鱼(sardine)

沙丁鱼在香港被人们称为沙甸鱼,又称萨丁鱼。小者长二寸(6.66 cm),大者尺(33.3 cm)许,下颚较上颚略长,齿不显,背苍腹白,肉美,多用来制成罐头食品。

沙丁鱼广泛分布于温带海洋中,产量极多,是世界上重要的海产经济鱼类,在西餐中的用量也很大。由于沙丁鱼体型小,产量多,因此其适宜做罐头。用沙丁鱼制作的罐头,肉质软烂,骨刺皆酥,在西餐冷菜中被大量使用。将沙丁鱼清蒸、红烧、油煎及腌干蒸食均味美可口。

新鲜的沙丁鱼鱼鳞无脱落,鱼眼清亮不混浊,鱼身紧,有弹性,鱼身两侧的黑斑非常明显。

Fresh sardines are often grilled, pickled or smoked, or preserved in cans, they are rich in vitamins and minerals.

(五)鱼子和鱼子酱(roe and caviar)

鱼子是由新鲜鱼子经盐水腌制而成,浆汁较少,呈颗粒状。鱼子酱是在鱼子的基础上经发酵而成的,其浆汁较多,呈半流质胶状。鱼子制品有红鱼子酱和黑鱼子酱两种。

1. 红鱼子酱

红鱼子酱是用大马哈鱼的卵制成的,制作方法:在鱼卵中加入4%的盐水,用木棍搅动,使衣膜与卵脱离,盐分渗入卵中,待腌透后滤出孵衣膜即红鱼子,如再发酵可制成红鱼子酱。

优质的红鱼子酱颗粒饱满、无破损、色红透亮、无汤汁,虽颗粒松散但附有少量黏液,味咸鲜,含盐率在4%以下。红鱼子酱常用作开胃小吃或装饰冷盘。

2. 黑鱼子酱

黑鱼子酱是用鲟鱼的卵制成的,因为鲟鱼的产量少,所以黑鱼子比红鱼子的价格贵。其加工方法与红鱼子相同。

优质的黑鱼子酱颗粒饱满,虽松散但有少量黏液,黑褐色有光泽,味清香鲜美,略带咸味,常用作开胃小吃或装饰冷盘。

Caviar is a delicacy which is consisted of salt-cured fish-eggs, it is considered as a delicacy and is eaten as a garnish or a spread.

Caviar is extremely perishable and must be kept refrigerated until its consumption. Pasteurized caviar has a slightly different texture, it is less perishable and may not require refrigeration before its opening. Pressed caviar is composed of damaged or fragile fish-eggs and can be a combination of several different roes which is specially treated, salted, and pressed.

二、其他水产品

(一) 龙虾(lobster)

龙虾又名大虾、龙头虾、虾魁、海虾等。原产地在中美洲、南美洲和墨西哥东北部地区。主要分布于热带海域,是名贵海产品。龙虾是世界上的大型食用虾类,肉洁白细嫩,味道鲜美,蛋白质含量高,脂肪含量低,营养丰富,倍受人们的青睐。龙虾是高档的烹饪原料,它既可用于做热菜,也可以用于做汤,适用于多种烹调方法,如煮、蒸、煎等,在法国菜、英国菜、美国菜中被经常使用。因为龙虾外壳美观,所以上菜时常常连同其甲壳一起上台,美观大方,饶有风味,经常用于宴会。

Lobster is used in soup, bisque, lobster rolls. Lobster meat may be dipped in clarified butter, resulting in a heightened flavour. Boil or steam live lobsters. When a lobster is cooked, its shell's colour will change from blue to orange.

(二) 明虾(shrimp)

明虾也称对虾,产于我国黄海、渤海及朝鲜西部沿海。其肉质细嫩,味道鲜美,在西餐中被大量使用,既可用于做汤,又能用于做菜。它适用于各种烹调方法,是非常理想的烹饪原料。常用于宴会,较为名贵。

Most shrimp are sold frozen and marketed based on their categorization of presentation, grading, colour, and uniformity.

Shrimp is rich in calcium, iodine and protein but low in food energy. Usually shrimp is sold whole, though sometimes only the meat of shrimp is marketed.

Shrimp roe, sometimes it is translated as Shrimp Flakes, is used as one of the ingredients in the preparation of sushi.

(三) 贻贝(mussel)

贻贝也叫青口,干制品则被称作淡菜。它是一种双壳类软体动物,壳呈黑褐色,生活在海滨岩石上。它分布于中国黄海、渤海沿岸。贻贝壳呈楔形,前端尖细,后端宽广而圆。

鲜活贻贝是大众化的海鲜品,由于贻贝产量大,收获后不易保存,因此人们历来多将其煮熟后加工成干品——淡菜。淡菜的营养价值很高,并具有一定的药用价值。贻贝的食用方法有多种,如清蒸、白灼、红焖、鲜炸、炒和煲汤等均可。

Mussels can be smoked, boiled, steamed, roasted, barbecued or fried in butter or vegetable oil. Mussels should be checked to ensure they are still alive just before they are cooked, a simple criterion is that live mussels will shut tightly when they are disturbed in the air. Open, unresponsive mussels are dead, and must be discarded. Unusually heavy, wild-caught, closed mussels may be discarded as they may contain only mud or sand. Mussel shells are usually open when cooked, revealing the cooked soft parts. Historically, it has been believed that after cooking all the mussels should have opened and those that have not are not safe to eat and should be discarded.

(四) 牡蛎(oyster)

牡蛎又称为蚝,目前大都为人工养殖,其肉质肥嫩鲜美,营养价值很高,是海产贝类中独具一格的原料,其肉多皱褶,外观上以蚝壳大而色深,相对较重者为佳。

牡蛎既可熟食又可生食,适宜炒、炸、烩或制汤等烹调方法,还可干制,或做成罐头。

Oysters must be eaten alive, or cooked alive. The shells of live oysters are usually tightly closed or snap shut when they are given a slight tap. If the shell is open, the oyster is dead, and cannot be eaten safely.

Oysters are commonly consumed by humans, cooked or raw, and they can be eaten raw, smoked, boiled, fried, stewed, canned, steamed. Eating can be as simple as opening the shell and eating the contents.

第五节 蔬 果 类

一、蔬菜类

西餐中蔬菜的种类较多,按照其食用部位可以将其分为叶菜类、根菜类、茎菜类、瓜菜类、豆菜类、花菜类等。

（一）蔬菜的分类

1. 叶菜类：食用部分主要为植物的叶子,多用来制作沙拉。

Leafy greens: spinach, beet greens, lettuces, endive, watercress.

Leaf vegetables are also called vegetable greens, leafy greens, or salad greens, sometimes accompanied by tender petioles and shoots. Although they come from a very wide variety of plants, most of them share a great deal with other leaf vegetables in nutrition and cooking methods.

（1）生菜(lettuce)

生菜按叶片的颜色分为绿生菜和紫生菜,按叶子的生长状态可分为散叶生菜和球生菜两种。生菜大都在沙拉的制作中使用,或单独使用,或和其他叶菜、肉类或芝士一起搭配。生菜还可以在汤、三明治的制作中使用,其茎部可以生食或烹制后食用。

Most lettuce is used in salads, either alone or with other vegetables, meats and cheeses. Lettuce leaves can also be found in soups, sandwiches, while the stems can be eaten both raw and cooked.

（2）西洋菜(watercress)

西洋菜俗称水田芥,为中空茎的水生植物,生长于流水、河塘及池塘中。它富含维生素C、维生素A及叶绿素、矿物质等,有刺激性的辣味和淡淡的苦香,在西餐中主要用来做配菜或切碎后制作调味汁等。

Watercress leaves go well with strongly flavoured meats such as game. The leaves are most commonly served raw as a garnish to eggs or meat, or as part of a salad with orange segments.

（3）菊苣（endive）

菊苣，为菊科多年生草本植物，国内外均有分布。菊苣为药食两用植物。菊苣叶片柔嫩多汁，营养丰富，氨基酸、维生素、胡萝卜素、钙的含量丰富。它的根含菊糖及芳香族物质，可提制代用咖啡，促进人体消化器官的活动。

菊苣适用于多种烹调方法，如凉拌、煎、烤等，在西餐中主要用来做沙拉原料。挑选时以外叶洁白有光泽、芽叶厚且抱合紧实者为佳，若叶菜前端出现绿色，则说明已不新鲜。菊苣在加热后，苦味会增强，因此清洗时不宜用热水冲洗，可在烹调时加入少许白糖来缓和其苦味。

Endive is a leaf vegetable belonging to the genus Cichorium, which includes several similar bitter leaf vegetables.

Endive is rich in many vitamins and minerals, especially in folate and vitamin A and vitamin K.

2. 花菜类：花菜类是以植物的花部为食用部位的蔬菜。常见的有西兰花、花菜等。

The cabbage family: cabbage, broccoli, cauliflower, brussels sprouts.

（1）西兰花（broccoli）

西兰花又名绿菜花、青花菜，属十字花科、芸薹属甘蓝的变种，其食用部分为绿色幼嫩花茎和花蕾，营养丰富，含蛋白质、脂肪、维生素，营养成分位居同类蔬菜之首，被誉为"蔬菜皇冠"。西兰花脆嫩爽口，风味鲜美、清香，在西餐中主要用于制作配菜和沙拉。

Broccoli is a green vegetable in the cabbage family, consisting of tight clusters of tiny green flower buds on fleshy stalks.

Wash well, soak the broccoli in salted water for 30 minutes if necessary to remove the insects, split large stalks into smaller sizes for portioning.

（2）抱子甘蓝（brussels sprouts）

抱子甘蓝又称芽甘蓝，其植株中心不生叶球，而在茎周叶腋中产生小芽球，犹如子附母怀，故称"抱子甘蓝"。其原产于地中海沿岸，19世纪初逐渐成为欧美国家的重要蔬菜之一。其以腋芽处形成的小叶球为食用部分，富含各种营养物质，适用于多种烹调方式，如清炒、素烧、凉拌、作汤料等。

The brussels sprouts is a nutritious and versatile vegetable, it is abundant in vitamin C, vitamin K and contains many antioxidants.

The most common method of preparing brussels sprouts for cooking begins with cutting the buds off the stalk. Any surplus stem is cut away, and any loose surface leaves are peeled and discarded. Once cut and cleaned, the buds are typically cooked by boiling, steaming, stir frying, grilling, slow cooking, or roasting. To ensure evenly cooking throughout, buds of a similar size are usually chosen. Some cooks will make a single cut or a cross in the center of the stem to aid the penetration of heat. Brussels sprouts can be pickled as an alternative to cooking.

Overcooking will make the buds gray and soft, for taste, roasting brussels sprouts is a common way to cook them to bring out flavor. Common toppings or additions for brussels sprouts include Parmesan cheese and butter, balsamic vinegar, apple cider vinegar, bacon, pine nuts, mustard, brown sugar, chestnuts. Another popular way of cooking brussels sprouts is to saute them.

(3) 甘蓝(cabbage(green, red and savoy))

甘蓝又称为洋白菜、卷心菜,学名结球甘蓝,是十字花科、芸薹属的植物。

甘蓝有青甘蓝、紫甘蓝和皱叶甘蓝三种。甘蓝的烹调方法比较多,最简单的方法是生吃,酸浸是保存甘蓝的一种常见方法(制成酸菜)。皱叶甘蓝一般在沙拉中使用。

Cabbages are large leaf clusters in dense, round heads. For both green and red cabbages, looking for a firm head, heavy size, good color, crisp leaves, finely ribbed is necessary. Savoy cabbages are not so heavy, with darker green, ruffled leaves.

Cabbage is prepared and consumed in many ways. The simplest option is eating the vegetable raw. Pickling is one of the most popular ways of preserving cabbage, creating dishes such as sauerkraut. Savoy cabbages are usually used in salads.

(4) 花椰菜(cauliflower)

花椰菜,又称花菜、菜花或椰菜花,为十字花科、芸薹属一年生植物。原产于地中海东部海岸,约在19世纪初清朝光绪年间引进中国。花椰菜,是一种很受人们欢迎的蔬菜,味道鲜美,营养丰富,维生素C的含量非常丰富,还具有抗癌功效,营养价值远远超出其他蔬菜。在烹调中,花椰菜可以烤、煮、煎、蒸。在烹调时,花椰菜外部的叶子和粗大的茎部一般会被去除,剩下菜花。

Cauliflower heads can be roasted, boiled, fried, steamed. When cooking, the outer leaves and thick stalks are typically removed, leaving only the florets. The leaves are also edible, but they are most often discarded. The florets should be broken into similar-sized pieces so that they are cooked evenly.

3. 茎菜类:食用部分为植物的茎部,西餐中常见的原料有洋葱、大蒜、葱等,嫩茎类的芦笋、竹笋、西芹等,块茎类的土豆、红薯等。

The stem of plant as edible part: onion, shallot, asparagus, celery, potato.

(1) 洋葱(onion)

洋葱是西餐中常见的蔬菜之一,其肉质柔嫩,汁多辣味淡,品质佳,适于生食。原产于亚洲西部,在中国各地均有栽培,四季都有供应。洋葱供食用的部位为生长在地下的肥大鳞茎(即葱头)。其在国外被誉为"菜中皇后",营养价值较高。洋葱的种类主要有:白洋葱、黄色洋葱、紫洋葱、珍珠状洋葱。

洋葱在热菜中的使用较多,如法式洋葱汤。此外,还可以烤、煮、焖、煎或在沙拉中直接食用。

Onion has a pungent, many-layered bulbous root, and it is kitchen's most commonly used vegetable. It comes in many varieties, from white, yellow, red and pearl onions.

Onions are commonly chopped and used as an ingredient in various warm dishes, for example in French onion soup. They are versatile and can be baked, boiled, braised, grilled, fried, roasted, or eaten raw in salads.

(2) 芦笋(asparagus)

芦笋是一种多年生的连作蔬菜,一次种植多年收益。其原产于欧洲地中海沿岸及小亚细亚地区,已有2000多年的种植历史,鸦片战争后,传教士将其带入中国。芦笋含有丰富的维生素B、维生素A以及叶酸、硒、铁、锰、锌等微量元素。芦笋含有人体所必需的各种氨基酸。

芦笋的口味清爽香郁,肉质细嫩,可以生吃凉拌或用来做配菜。

Asparagus is served in a number of ways around the world, typically as an appetizer or vegetable side dish.

In Asian-style cooking, asparagus is often stir-fried.

(3) 西芹(celery)

西芹又称洋芹,是从欧洲引进的芹菜品种,植株紧凑粗大,叶柄宽厚、实心。其质地脆嫩,有芳香气味。西芹的营养丰富,茎叶中含有挥发性芳香油,可促进食欲。在西餐中,西芹的使用广泛,可制作成沙拉凉拌生吃,也可炖煮、腌制原料、捣汁加热饮用。

Celery has a long fibrous stalk. Depending on location and cultivar, its stalks, leaves are eaten and used in cooking. Celery seed is also used as a spice, its extract is used in medicines.

Celery is eaten as a vegetable around the world. In Europe, the hypocotyl is used as a root vegetable. The leaves are strongly flavoured and are not commonly used, either as a flavouring in soups or as dried herb. Celery is a staple in many soups, such as chicken noodle soup.

4. 根菜类:根菜类的食用部分为植物的根部。西餐中常见的原料有胡萝卜、芜菁、红菜头等,通常生长于土中,烹调之前将其洗净去皮,适用于多种烹调方法。

The root of plant as edible part: turnip, beetroot.

(1) 芜菁(turnip)

芜菁,别名蔓菁、圆根、盘菜。芜菁的肥大肉质根供食用,肉质根柔嫩、致密,可供炒食、煮食。芜菁的根以及叶子都可食用,其富含维生素A、叶酸、维生素C、维生素K和钙。

The turnip is a root vegetable commonly grown in temperate climates worldwide.

Small, tender varieties are grown for human consumption, while larger varieties are grown as feed for livestock.

In the United States, stewed turnips are eaten as a root vegetable in the autumn and winter. The greens of the turnip are harvested and eaten all year. Turnip greens may be cooked with a ham hock or pieces of fat pork meat, stewed turnip greens are often eaten with vinegar.

（2）红菜头(beetroot)

红菜头又称为火焰菜,是红色根用甜菜,主要以球形的肉质根供人们食用,是欧美各国的主要蔬菜之一,原产于地中海沿岸,有2000多年的栽培历史。可食用部位为其肥大的肉质根,含纤维少,质地致密柔嫩,适用于煮食或炒食。可配牛肉同炒,或作汤料,也可凉拌、作配菜,其颜色艳丽,诱人食欲。

Usually beetroot is eaten boiled, roasted or raw, and either alone or combined with any salad vegetable.

Beetroot can be boiled or steamed, peeled and then eaten warm with or without butter as a delicacy; cooked, pickled, and then eaten cold as a condiment; or peeled, shredded raw, and then eaten as a salad.

5. 瓜菜类:瓜菜类是以植物的瓜果为烹饪原料的蔬菜,在西餐中常见的有节瓜、南瓜、黄瓜等。

Gourd vegetable: zucchini, pumpkin.

（1）节瓜(zucchini/summer squash)

节瓜是一种夏季产的南瓜,也称西葫芦,其颜色可以是深绿色或浅绿色,在烹饪中,可作为开胃菜或配菜。一般经烹调后食用,其烹调方法多样,包括蒸、煮、烤、填馅、煎等。

Zucchini is a summer squash, it can be dark or light green. In a culinary context, it is treated as a vegetable. It is usually cooked and presented as a savory dish or accompaniment.

Zucchini is usually served cooked. It can be prepared by using a variety of cooking techniques, including steamed, boiled, grilled, stuffed and baked, barbecued, fried.

（2）南瓜(pumpkin)

南瓜在烹饪中的用途多样,其外皮、籽、叶子和花都可以食用。成熟后的南瓜的烹调方法多样,包括煮、蒸、烤。

在北美洲,南瓜是一种传统的节日食物,可将其捣成泥状或制成汤,也可以做成感恩节的重要食物——南瓜派。

Pumpkin is a kind of winter squash that is round, with smooth, slightly ribbed skin, and deep yellow to orange coloration.

Pumpkins are very versatile in their uses for cooking. Most parts of the pumpkin are edible, including the fleshy shell, the seeds, the leaves, and even the flowers. When ripe, the pumpkin can be boiled, steamed or roasted.

In North America, pumpkin is a very important, traditional part of the autumn harvest, it is eaten mashed and made into soups and purees, it is often made into pie.

6. 种子和豆荚类：食用部分为植物的种子部分，西餐中常见的品种有：荷兰豆、秋葵等。

Seeds and pods: beans, peas, corn, okra.

（1）荷兰豆(sweet broad pea)

荷兰豆是豆科豌豆属。最早栽培荷兰豆的地区是大约 12000 年前的泰缅边境地带。在 17 世纪，荷兰人从世界各地带来各种舶来品，于是，当地居民开始称之为荷兰豆。

荷兰豆的营养价值丰富，口感鲜嫩、清香，在西餐中主要用于做配菜等。

（2）秋葵(okra)

秋葵的原产地为印度，广泛栽培于热带和亚热带地区，素有"蔬菜王"之称，有极高的经济用途和食用价值，其嫩荚肉质柔嫩，含有由果胶及多糖组成的黏性物质，有一种特殊的风味，口感爽滑。一般可用来炒食、做汤、腌渍、罐藏等，除嫩果可食外，其叶片、芽、花也可食用。

Okra is valued for its edible green seed pods, pods are eaten raw, cooked, pickled, or included in salads.

（二）处理新鲜蔬菜的方法

1. 清洗(washing)

（1）Wash all the vegetables thoroughly.

所有的蔬菜需要彻底清洗。

（2）Root vegetables that are not peeled, such as potatoes for baking, should be scrubbed very well with a stiff vegetable brush.

不需要去皮的根类蔬菜，如用于烘烤的土豆，其外皮应用坚硬的刷子擦洗干净。

（3）Wash green, leafy vegetables in several changes of cold water. Lift the greens out from the water so the sand can sink to the bottom.

绿叶蔬菜需要进行多次清洗，待其干净后沥干。

（4）After washing, drain well and refrigerate lightly. The purpose of covering is to prevent drying, but if the covering is too tightly, it will cut off air circulation. This can be a problem if the product is stored more than a day, because mold is more likely to grow in a damp, closed space.

将蔬菜洗净后沥干。为了防止蔬菜变干，应将其盖住，但不宜盖得过紧，否则影响空气流通。若蔬菜的保存时间在一天以上将会出现问题，因为霉菌更有可能在潮湿、密闭

2. 浸泡(soaking)

(1) Do not soak vegetables for a long time, or the flavor and nutrients will leach out.

避免长时间浸泡蔬菜,以免香味和营养流失。

(2) Cabbage, broccoli, brussels sprouts, and cauliflower may be soaked for 30 minutes in the cold salted water to eliminate insects, if necessary.

如有必要,卷心菜、西兰花、花椰菜等应在冷盐水中浸泡30分钟以去除昆虫。

(3) Limp vegetables can be soaked briefly in the cold water to restore their crispness.

柔软的蔬菜可以在冷水中简单浸泡,以恢复其脆度。

3. 去皮和切配(peeling and cutting)

(1) Peel most vegetables as thinly as possible. Many nutrients lie just under the skin.

蔬菜去皮时,尽量削得薄些,因为很多营养素在外皮下面。

(2) Cut vegetables into uniform pieces for even cooking.

切配蔬菜时使其大小一致便于烹调。

(3) Peel fruits and vegetables as close to cooking time as possible to prevent drying and loss of vitamins through oxidation.

水果和蔬菜去皮后要尽快烹调,防止由于氧化而导致的脱水和维生素流失。

(4) Different vegetables can be cut into uniform size to minimize waste by using a paring machine.

使用削皮机可以将不同的蔬菜分类切割成大小均匀的形状,以减少浪费。

(5) Treat vegetables that brown easily(potatoes, eggplant, artichokes, sweet potatoes) with an acid such as lemon juice, until they are ready to use.

用柠檬汁等酸性物质处理一些容易褐变的蔬菜(如土豆,茄子,朝鲜蓟,红薯),以备使用。

(三) 蔬菜烹饪的基本原则

1. Don't overcook.

不要过度烹调。

2. Cook as close to service time as possible, and in small quantities. Avoid holding vegetables for long periods of time on a steam table.

菜品烹制好后应立即端给客人,菜品要少量烹饪。避免将菜品长时间放置在食品台上保温。

3. If the vegetables must be cooked ahead, undercook slightly and chill rapidly. Reheat them at the service time.

如果蔬菜需要提前烹制,应缓慢烹调、快速冷却。上菜时预热菜品。

4. Never use baking soda with green vegetables.

不要把苏打粉和绿叶蔬菜一起使用。

5. Cut vegetables uniformly for even cooking.

切蔬菜时保证蔬菜大小一致,以便于烹调时受热均匀。

6. Cook green vegetables and strong-flavored vegetables uncovered.

烹饪绿叶蔬菜和气味浓烈的蔬菜时,不需要加盖。

7. To preserve color, cook red and white vegetables in the slight acid(not strong acid) liquid. Cook green vegetables in a neutral liquid.

为了保持蔬菜的颜色,应在弱酸性的液体中烹调红色和白色的蔬菜,在中性液体中烹调绿叶蔬菜。

8. Do not mix a batch of freshly cooked vegetables with a batch of the same vegetables that were cooked earlier and kept hot in a steam table.

不要将不同时间段烹饪的同一种蔬菜混合或一起放置在具有蒸汽保温设备的餐桌上保温。

二、果品类

果品类原料有着丰富多样的颜色,在西餐中广泛应用,可以用来制作沙拉、甜品。西餐中常见的果品类原料有:橙子、苹果、葡萄、香蕉、西瓜、柠檬、菠萝、草莓、猕猴桃、哈密瓜、青柠、牛油果、鳄梨、蓝莓、芒果、西柚等。

1. 柠檬(lemon)

柠檬又名柠果、洋柠檬、益母果,原产地是地中海沿岸及我国云南地区,其果形优美,颜色鲜艳,惹人喜爱。柠檬含有丰富的柠檬酸、柠檬香精油、类黄酮、维生素 C、维生素 A、维生素 P 以及钙、铁、锌等多种微量元素,还含有苹果酸、葡萄酸、琥珀酸等十多种有机酸。其味极酸,微苦,因此不能像其他水果一样生吃鲜食。在西餐桌上它却是不可少的珍品,故被誉为"西餐之宝"。在西餐中,无论是冷菜、热菜、甜品,还是饮品都离不开用柠檬调味。

Lemon juice, rind, and zest are used in a wide variety of foods and drinks. Lemon juice is used to make lemonade, soft drinks, and cocktails. It is also used in marinade for fish.

Lemon juice is also used as a short-term preservative on certain foods that tend to oxidize and turn brown after being sliced (enzymatic browning), such as apples, bananas, and avocados.

Lemon juice and rind are used to make marmalade, lemon curd and lemon liqueur. Lemon slices and lemon rind are used as a garnish for foods and drinks. Lemon zest, the grated outer rind of the fruit, is used to add flavor to baked goods, puddings, rice and other dishes.

The leaves of the lemon tree are used to make tea and for preparing cooked meats and seafoods.

2. 青柠(lime)

青柠也称绿檬,其果实呈淡黄绿色的球形、椭球形或倒卵形。由于亚热带与热带地区出产的柠檬果皮也呈现绿色,因此青柠常被这些地区的民众误认为柠檬。

青柠檬与黄柠檬相比,青柠檬没有那么酸,但其气味浓烈,口味清甜,主要用于搭配东南亚菜系中辛辣、酸辣的味道。在鸡尾酒或冷饮中,通常可将青柠切片用于装饰或泡在饮料、酒之中起到提味的作用。

3. 牛油果(avocado)

牛油果又名油梨、鳄梨、酷梨、奶油果,有"森林黄油"之美称,是一种营养丰富的水果,含多种维生素和丰富的脂肪酸,还含有蛋白质及钠、钾、镁、钙等元素,一般作为生果食用,也可被制作为菜肴和罐头。

Avocado has higher content of fat than most other fruits. It can be served as an important staple in the diet of consumers who have limited access to other fatty foods (high-fat meats and fish, dairy products).

Avocado is used in both savory and sweet dishes, it is popular in vegetarian cuisine as a substitute for meats in sandwiches and salads because of its high fat content.

Generally, avocado is served raw, frequently used for milkshake and occasionally added to ice cream and other desserts.

4. 西柚(grapefruit)

西柚成熟时果皮一般呈不均匀的橙色或红色,果肉呈淡红白色。进口西柚的主要产地包括:南非、以色列、中国台湾。西柚以重量相当,果身光泽皮薄、柔软的为好。西柚富含维生素 C,含糖较少。西柚有多种吃法,除了直接食用外,还可榨成汁拌沙拉和凉菜。做海鲜时加点西柚汁能起到去除腥味的作用。

The grapefruit is known for its sour to semi-sweet, it is rich in vitamin C. It is used in salads and seafood dishes.

5. 蓝莓(blueberry)

蓝莓意为蓝色浆果,属杜鹃花科,越橘属植物。其起源于北美,为多年生灌木小浆果果树。因果实呈蓝色,故称为蓝莓。蓝莓果肉细腻,风味独特,酸甜适度,又具有香爽宜人的香气,蛋白质、维生素等常规营养成分的含量十分丰富,矿物质和微量元素的含量也相当可观,因此,蓝莓被誉为"水果皇后"。

Blueberries are sold fresh or are processed as individually quick frozen fruit, juice, or dried. Then these may be used in a variety of consumer goods, such as jams, blueberry pies, muffins, snack foods.

Blueberry wine is made from the flesh and skin of the berry, which is fermented and then matured.

第六节 奶 类

一、牛奶

牛奶也称牛乳,营养价值非常高,含有丰富的蛋白质、脂肪及多种维生素和矿物质,经消毒处理的新鲜牛奶有全脂、半脱脂和脱脂三种类型。

Whole milk(全脂奶) is fresh milk as it comes from the cow, with nothing removed and nothing(except vitamin D)added. It contains about 3.5% fat, 8.5% nonfat milk solids and 88% water.

Skim med or nonfat milk(脱脂奶) has had most or all of the fat removed. Its fat content is usually indicated, usually 1% or 2%.

Fortified nonfat or low-fat milk(低脂奶) has had substances added to increase its nutritional value, usually vitamins A and D and extra nonfat milk solids.

Flavored milks(调味奶), such as chocolate milk, have had flavoring ingredients added. A label such as chocolate milk drink or chocolate-flavored drink indicates the product does not meet the standards for regular milk.

新鲜牛奶应为乳白色或略带浅黄色,无凝块,无杂质,有乳香味,清新自然,品尝起来略带甜味,无酸味。

牛奶的保存一般采用冷藏法。长期保存应放在$-18 \sim -10\ ℃$的环境下,短期保存应放在$-2 \sim -1\ ℃$的冰箱中。

二、奶油

奶油是从高温杀菌的鲜乳中,经过加工而分离出来的脂肪和其他成分的混合物,在乳品工业也称稀奶油。奶油是制作黄油的中间产品,含脂率较低,一般为20%~25%。

奶油为乳白色，略带浅黄色，呈半流质状态，在低温下较稠，经加热可熔为可流动的液体。优质的奶油气味芳香纯正，味稍甜，组织细腻，无杂质，无结块。品质较差的奶油有异味，并含有奶团杂物。

奶油的保管一般采取冷藏法，保存温度为 4～6 ℃。为了防止污染，无包装的奶油应放在干净的容器内，并加上盖。由于奶油的营养丰富，水分充足，极易变质，因此要注意及时冷藏。其制品在常温下超过 24 h 不应食用。

（一）新鲜的奶油制品（fresh cream products）

（1）Heavy cream is also called double cream, it contains 48 to 50 percent fat.

厚奶油的乳脂含量为 48%～50%。这种奶油的用途不广，因成本过高，通常情况下为了增进风味才使用。

（2）Whipping cream has a fat content of 30 to 40 percent.

掼奶油的乳脂含量为 30%～40%，很容易被搅拌成泡沫状，主要用于裱花装饰。

（3）Light cream is also called table cream or coffee cream, it contains 18 to 30 percent fat.

淡奶油的乳脂含量为 18%～30%，具有起稠增白的作用，可用于少司的调味或用来做西点的配料。

（二）发酵牛奶及奶油制品（fermented milk and cream products）

Sour cream has been fermented by added lactic acid bacteria, which makes it thick and slightly tangy in flavor. It has about 18 percent fat.

酸奶油是由添加的乳酸菌发酵制成的，这使得酸奶油变稠，气味变得浓烈。酸奶油的脂肪含量约为 18%。

Yogurt is milk cultured by special bacteria. It has a custard-like consistency.

酸奶是牛奶经过特殊菌种发酵而成，其浓度和蛋奶糊很相似。

（三）奶油的打发（whipping cream）

Cream with a fat content of 30 percent or more can be whipped into a foam. For the best results, please observe the following guidelines.

（1）Cream and all equipments are well chilled.

奶油和打发设备需要冷藏。

（2）Do not sweeten until the cream is whipped. Sugar decreases the stability of cream and makes it harder to whip.

打发过程中不要加糖。糖可以降低奶油的稳定性从而使打发难度加大。

（3）Do not overwhip. Stop beating when the cream forms stiff peaks.

不要过度搅打。

三、黄油

黄油(butter)是从奶油中进一步分离出来的脂肪,分为鲜黄油和清黄油两种。鲜黄油含脂率在85%左右,口味香醇,可以直接食用。清黄油含脂率在97%左右,比较耐高温,可用于烹调热菜。

黄油在常温下为浅黄色固体,加热后融化,并有明显的乳香味。黄油具有良好的起酥性、乳化性和可塑性,但其储存稳定性较差。

黄油脂肪率较高,如长期储存应放在-10 ℃的冰箱中,短期保存可放在5 ℃左右的冰箱中冷藏。存放时应避免光线直接照射,且密封保存,避免黄油氧化。

Margarine is a manufactured product, it is made from vegetables and animal fats, plus flavoring ingredients, coloring agents and so on.

人造黄油在口感、外形上和黄油相似,主要是利用动植物的脂肪、调味料、色素等制作而成,和黄油一样,其脂肪含量约为80%。

四、奶酪

奶酪是一种发酵的牛奶制品,其性质与常见的酸牛奶有相似之处,都是通过发酵过程来制作的,也都含有可以保健的乳酸菌,但是奶酪的浓度比酸奶更高,近似固体食物,营养价值也更加丰富。1 kg的奶酪制品是由10 kg的牛奶浓缩而成的,其含有丰富的蛋白质、钙、脂肪、磷和维生素等营养成分,是纯天然的食品。

Cheese is zymolytic dairy product, the character is similar as normal yogurt, they are both made from zymolysis, but the cheese contains more lactobacillus, so it is more nutritious. Each 1 kg cheese is made form 10 kg of milk, it contains abundant protein, fat, vitamins, calcium, it is natural food.

奶酪的起源,最普遍的说法认为它是由游牧民族发明的。他们早先将鲜牛奶存放在牛皮背囊中,但往往几天后牛奶就会发酵变酸。后来他们发现,变酸的牛奶在凉爽湿润的气候下经过数日,会结成块状,变成极好吃的奶酪,于是这种保存牛奶的方法得以流传,奶酪也一直是这些游牧民族的主要食物之一。虽然奶酪比较耐储藏,但奶酪其实始终处于发酵过程中,因此时间太长了也会变质。

(一)奶酪的分类

奶酪是西餐中必不可少的菜肴。目前,全世界约有数千种以上的天然奶酪,产地涵盖将近整个欧洲,以及美国、澳洲、新西兰与日本等地。其中,每个国家都有其代表性的奶酪,如意大利的 Gorgonzola、Mascarpone、Ricotta,荷兰的 Gouda、Edam,英国的 Stilton、Cheddar,瑞士的 Emmenthal、Gruyere,德国的 Bonifaz、Bavariablu,丹麦的 Danadlu、Havarti。

奶酪可分为加工奶酪与天然奶酪。事实上,我们最常用来夹面包、做三明治以及超市货架上随处可见的,都是加工奶酪。在欧洲地区,天然奶酪才是奶酪中的主流,和葡萄酒一样,大多数天然奶酪有悠久的历史传承,至今仍大半保留了传统的手工制作精神,并随着产区、气候、土壤、牧草、海拔高度、制作过程、配方与熟成程度的不同,而拥有千变万化的口感风貌。天然奶酪的制作,主要步骤为先用乳酸菌和酵素使牛奶凝结,然后经过切割、搅拌、去除乳清与水分的过程后,填装于模型内,再压榨、加盐,等待其熟成。天然奶酪依照个别的熟成方式与硬度,主要可分为以下几种。

1. 新鲜奶酪(fresh cheese)

These are soft, white, freshly made cheese.

不经熟成,直接将牛乳凝固之后,去除部分水分而成的新鲜奶酪,呈现出洁白的颜色与柔软湿润的质感,散发出清新的奶香与淡淡的酸味,十分爽口。例如:奶酪蛋糕的主要原料 cream cheese 来自意大利,经常使用于沙拉或开胃菜中的 Mozzarella;常用以制作 Tiramisu 的 Mascarpone、低脂清爽的 Ricotta,以及产自法国、以著名美食家为名的 Brillat Savarin 等,都是人们极为熟悉的新鲜奶酪。

(1) 奶油奶酪:Cream cheese is a smooth, mild cheese with a high fat content. It is extensively used in making sandwiches, starters and in baking.

奶油奶酪的脂肪含量较高,主要用于烘焙、三明治和头盘菜肴的制作。

(2) 马苏里拉奶酪:Mozzarella is a soft, mild cheese made from whole milk or part skimmed milk. It has a stringy texture that comes from being pulled and stretched during its production. It is widely used in pizzas and Italian-style dishes.

马苏里拉奶酪主要是由全脂奶或部分脱脂奶制成,主要用于比萨和意大利菜肴的制作。

(3) 马斯卡彭奶酪(Mascarpone cheese):如图 2-1(彩图 1)所示。其脂肪含量为 75%,是一种未经过发酵的鲜奶酪,质地细腻,味道温和。其口感浓郁滑顺,略带酸味。这种干酪不是用牛奶或凝乳制成的,而是用奶油经过再精炼制成的。它是制作著名的甜点提拉米苏的重要原料,也可加入果酱或与水果一同食用。

图 2-1

马苏里拉蚕豆沙拉(Mozzarella with broad bean

salad) （图 2-2(彩图 2)）

图 2-2

 原料 Ingredients

300 g podded fresh or frozen broad beans	300 g 新鲜有荚的或冷冻的蚕豆
100 g salad leaves	100 g 沙拉叶子(野蒜叶,豌豆苗和水芹)
(wild garlic leaves, pea shoots and cress)	
500 g Mozzarella or feta	500 g 马苏里拉奶酪,掰成一口大小
(broken into rough, bite-size pieces)	

 调味汁 For the dressing

2 garlic cloves, crushed	2 瓣大蒜,捣碎
juice of 1 lemon	1 个柠檬的汁
5 tablespoons of olive oil	5 汤匙橄榄油
a little chopped fresh mint	少量切碎的新鲜薄荷叶

 工具 Tools

stockpot	汤锅
wooden spatula	木铲

 制作方法 Method

(1) Blanch the beans in plenty of salted, rapidly boiling water for no more than a couple of minutes. Plunge the beans straight into cold water and drain. Squeeze the beans gently with your thumb and forefinger to remove the tough skins.

在含盐的沸水中烫煮蚕豆不超过 2 min,后直接用冷水冲洗、沥干,用拇指和食指轻轻挤豆子以去除其外壳。

（2） Make the dressing, mix the garlic, lemon juice, oil, mint and some salt together, then set it aside.

制作调味汁,将大蒜、柠檬汁、油、薄荷叶和盐混合,后静置。

（3） Arrange the salad leaves, beans and cheese on plate, and serve with the dressing.

摆盘,将沙拉叶、豆子、奶酪放入盘中,和调味汁一起端给客人。

 特点　Characters

Light and refreshing　　　　　　　　　　清淡爽口

2. 白霉奶酪(white mould cheese)(图2-3(彩图3))

表面上覆盖着一层白霉为其主要特征。当霉菌在表面繁殖发酵时,奶酪内部也会随之渐渐熟成;而因为白霉的作用,这类奶酪的质地十分柔软,尤其是已达完全熟成状态的,更是浓稠滑腻、奶香浓郁、口感独特。

其中以村庄为名,深受拿破仑三世喜爱,散发着浓郁的牛乳与奶油香气的法国诺曼底Camembert最具代表性;再如曾获得世界一等评价,口感芬芳甜美的Brie de Meaux,口感润泽、带有微微坚果香的Colommiers,以及浪漫心形的Neufchatel也十分知名。

Brie cheese(布里奶酪):由牛乳制成,其外形最有特色的是表皮有细细的条状的纹路,比较白,内里则有少少泛黄,口感浓郁绵密,脂肪含量一般为45%。

Camembert cheese(卡蒙贝尔奶酪):脂肪含量同布

图2-3

里奶酪差不多,奶香味十足,一般都是由牛乳制成。可以搭配开胃菜,水果蔬菜等。

Coulommiers cheese(库洛米耶尔奶酪):外形与布里奶酪很相似,但是其体型较小,熟成的时间也更长一点。由牛乳制成。它上面的菌落是青黴菌,脂肪含量大致为40%,口感也非常浓郁。

3. 蓝纹奶酪(blue cheese)(图2-4(彩图4))

蓝纹奶酪是所有奶酪中风味最特别的一种,其制作方法:将蓝霉与凝乳均匀混合后,一起填装于模型中进行熟成。其组织中布满着如大理石纹般美丽的蓝色纹路,滋味也大有别于温和的白霉奶酪,有着强劲刺激、辛香浓烈的风味,个性十足。其中,来自法国,据说已有2000年历史的Roquefort,英国的Stilton,以及意大利的Gorgonzola,并称为世界三大蓝纹奶酪。

These cheese own their flavour and appearance to the blue or green mold that mottles their interiors. The most famous of the blue cheeses is Roquefort, made in France from sheep's milk. Stilton, from England, is a mellower, firmer blue cheese. Italy's gorgonzola is a soft, creamy cheese with an unmistakable pungency.

图 2-4

4. 洗浸奶酪(wash rind cheese)(图 2-5(彩图 5))

洗浸奶酪是利用细菌进行熟成的奶酪,在熟成期间以盐水或当地特产酒再三擦洗表皮,使之渐渐产生馥郁强烈的香气与黏稠醇厚的口感,尤其是经过当地酒擦洗制成的奶酪,往往有着浓厚的地域气息,格外迷人。

These cheeses are ripen from the outside toward the center. When they are young, they are firm and cakey and have little flavor. As they mature, they gradually become soft. The ripening starts from the inside of the rind and spreads to the center.

图 2-5

5. 半硬奶酪(semihard cheese)(图 2-6(彩图 6))

半硬奶酪是在制造过程中强力加压、去除部分水分后所形成的。由于其口感温和,因而最容易被一般人所接受与喜爱。因为其易于溶解,所以也常被大量用于菜肴的烹调中。其中,以生产于法国阿尔卑斯山区、奶香浓郁的 Reblochon 与 Tomme de Savoie,

以及产自荷兰、风味平易近人的 Gouda 等最具知名度。

Cheeses that range in texture from semi-soft to firm include Swiss-style cheeses such as Emmental and Gruyère. The same bacteria that contribute to their aromatic and sharp flavours. Other semi-soft to firm cheeses include Gouda, Edam, Jarlsberg, and Cantal. Cheeses of this type are ideal for melting and are often served on toast for quick snacks or simple meals.

图 2-6

6. 硬质奶酪（hard cheese）（图 2-7（彩图 7））

质地坚硬、体积硕大、沉重的硬质奶酪，是经过至少半年甚至两年以上而熟成的奶酪。它不仅可耐长时间的运送与保存，且经久酝酿而散发出浓醇甘美的香气，十分耐人寻味。

Hard cheeses have lower moisture content than softer cheeses. They are generally packed into moulds under more pressure and aged for longer time than soft cheeses.

图 2-7

Hard cheeses (grating cheeses such as Parmesan and Pecorino Romano) are quite firmly packed into large forms and aged for months or years.

（二）奶酪与酒的搭配

比较清淡的新鲜奶酪与半硬奶酪可搭配口味清淡、带有果香的红酒或是较干型白酒；口感黏稠醇厚的山羊奶酪与白霉奶酪可搭配浓郁的红酒；个性十足的蓝纹奶酪可搭配强劲浓厚的红酒或 Muscat、Sauternes 等甜白酒。

咸度高的奶酪与略酸的葡萄酒，脂肪含量多的奶酪与较干型的葡萄酒，由于彼此具有中和协调的作用，因此同样十分相配。

奶酪与葡萄酒都具有极强的地域特性与风格，尤其法国的奶酪产区与葡萄酒产区分布呈现大致重叠的状况。

因此，产地相同或相似的奶酪与红酒，如阿尔萨斯的 Munster 搭配阿尔萨斯的白酒，勃艮第的 Epoisse 可搭配勃艮第红酒或马尔酒，诺曼底的 Camenbert 则搭配当地的苹果酒。

（三）奶酪的保存与服务

1. 奶酪的保存(cheese storing)

In general, the firmer and more aged cheese is, the longer it will keep.

不同的奶酪其保存方法也不尽相同。一般情况下，越是坚硬的、发酵时间越长的奶酪，其保存时间越长。

2. 奶酪的服务(cheese serving)

Serve cheese at room temperature. This is the single most important rule of cheese service. Only at room temperature will develop the full flavor of cheese. (This does not apply to unripened cheese like cottage cheese.)

Cut cheese just before service to prevent drying.

奶酪的服务之所以要在室温下进行，是因为只有在室温下，奶酪的香味才能完全释放出来（由脱脂凝乳制成的松软干酪除外）。在服务前切奶酪能够更好地防止奶酪变干。

（四）烹调奶酪时需要注意的事项

（1）低温烹调。

Use low temperature. Cheese contains high content of protein, which toughens and becomes stringy when heated too much. Sauces containing cheese should not be boiled.

（2）较短的烹调时间。

Use short cooking time, for the same reason. Cheese should be added to the sauce at the end of cooking.

（3）保持磨碎的奶酪大小一致。

Grate cheese for faster and more uniform melting.

（4）年限时间长的奶酪比时间短的奶酪更容易融化和与食物混合。

Aged cheeses melt and blend into foods more easily than young cheeses.

（5）年限时间长的奶酪比时间短的奶酪更具风味，烹调时所需的量要少。

Aged cheeses add more flavor to foods than young, mild cheeses, so you need less of it.

五、炼乳

炼乳是将鲜乳经真空浓缩或其他方法除去大部分的水分，浓缩至原体积 25% ～

40%的乳制品。

炼乳加工时根据所用的原料和添加的辅料不同,可以分为加糖炼乳(甜炼乳)、淡炼乳、脱脂炼乳、半脱脂炼乳、花色炼乳、强化炼乳和调制炼乳等。炼乳多以罐装形式存放,一般采用冷藏法。

第七节 西餐调味品

一、食盐(salt)

食盐是对人类生存最重要的物质之一,也是烹饪中最常用的调味料。盐的主要化学成分氯化钠(化学式 NaCl)在食盐中的含量为 99%(属于混合物)。按食盐的来源可分为海盐、湖盐、井盐和岩盐,在西餐烹饪中以海盐的使用最为普遍,海盐按加工方法的不同又可分为大盐、精盐。

(一) 大盐

大盐是在沿海地区利用自然条件把海水晒制成饱和溶液,使氯化钠结晶析出而形成的。大盐颗粒大,结构紧密,色泽灰白,氯化钠含量在 94% 左右。由于其颗粒大,溶解慢,略带苦涩味,因此不适合用于在烹调中调味,但适宜用于腌制菜肴。

(二) 精盐

精盐也称为再制盐,是将大盐溶化成饱和溶液后,去除杂质,再经蒸发而成的。精盐呈粉末状态,色洁白,质地纯,氯化钠含量在 96% 以上,溶解快,适宜调味。

优质的食盐色泽白,味纯正,结晶小,疏松,不结块。食盐易溶于水,吸湿性强,若环境湿度超过 70%,食盐就会潮解,因此食盐应保存在干燥的容器内,并注意保持清洁卫生。

二、食糖(sugar)

食糖是用甘蔗或甜菜作为原料,经榨汁后加工制成的调味品。食糖种类较多,西餐中常用的有以下品种。

（一）白砂糖（granulated sugar）

白砂糖是食糖中最纯的一种，食糖含量在99％以上，色泽洁白，颗粒为结晶状，甜味纯正，在西餐烹调中使用广泛。

（二）绵白糖（white soft sugar）

绵白糖的食糖含量为97％～98％，含有少量水分和还原糖，质地细腻，溶解快，适合制作快速烹调的菜肴。

（三）红糖（brown sugar）

红糖是未经提纯的甘蔗制品。红糖色褐红，光亮，绵软，带有甘蔗的香味，适合制作圣诞布丁等甜品。

（四）方糖（cubic sugar）

使用优质砂糖加工压制而成，长方形，色洁白，主要放在西餐餐台上，可直接用于咖啡、红茶等饮料。

三、醋（vinegar）

醋是西餐中常见的调味品之一，以下是西餐中常用的醋的种类。

（一）苹果醋（apple vinegar）

通常先将苹果中的糖转化成酒精，然后再将酒精转化成醋酸，此为苹果醋。自制的苹果醋很混浊，但市场上出售的苹果醋却很清澈，主要原因是市场上出售的苹果醋经过了蒸馏。在烹调上，苹果醋的用途广泛，可替代米醋使用，也可以当作沙拉酱使用。

（二）白醋（white vinegar）

白醋或蒸馏醋是经过蒸馏和提纯的，因此酸味比较纯正。

（三）酒醋（grape vinegar）

酒醋是用葡萄或酿葡萄酒的糟渣发酵而成，有红葡萄醋和白葡萄醋两种。品质最好的酒醋是在橡木桶里慢慢酿造出来的，因为缓慢处理能使其香味更为集中。白葡萄醋最适合用来调制马乃司或类似的少司。

雪利酒醋由雪利葡萄酒制成，因而具有雪利酒特有的味道。

加味醋的醋汁是酒精，通常浸有香料，包括罗勒、百里香、糖、盐、月桂叶和迷迭香。

（四）其他特殊的醋

其他特殊的醋包括麦芽醋、米醋、果味醋等。醋的酸度决定了由醋制成的少司的味道，大多数少司的醋酸度在5%，但有些少司的醋酸度在7%～8%，酸性太强的醋在按照食谱使用时应加水稀释后再使用。酒醋通常被用来做最好的油醋汁。

四、番茄酱和番茄少司（ketchup and tomato sauce）

番茄酱和番茄少司是西餐中广泛使用的调味品。

番茄酱（ketchup）是鲜番茄的酱状浓缩制品。酱体呈鲜红色，浓度适中，质地细腻，无颗粒，无杂质。番茄酱常用作鱼、肉等食物的烹饪佐料，是增色、添酸、助鲜、提香的调味佳品。

番茄少司（tomato sauce）是一种以番茄为主要原料，辅以各种其他调味料制成的酱料。一般将其作为制作肉食和蔬菜的酱料，但最常见于制作意大利面等食品时作为调料。番茄少司有多种口味，最常见的有大蒜口味、甜椒口味、辣椒口味、海鲜口味等。

番茄少司和番茄酱最大的区别在于，番茄酱可以直接食用，而番茄少司必须经过烹饪处理才能食用。

五、辣椒汁和辣酱油（Tabasco and Worcestershire sauce）

辣椒汁（Tabasco），由美国艾弗瑞岛的麦克尼家族创制，至今已有130多年历史，已推出了四款不同口味的辣椒汁。

辣酱油，是一种英国调味品，又称喼汁、辣醋酱油、英国黑醋或伍斯特少司（Worcestershire sauce）。李派林喼汁是辣酱油的典型代表，自1838年发售，由于其发明和最早生产地点是Lea & Perrins在伍斯特郡的郡府伍斯特的作坊，因而命名为"伍斯特少司"。

辣酱油在19世纪初传入我国，因其色泽风味与酱油相似，所以习惯上称为"辣酱油"。

辣酱油与普通酱油稍有不同，其味道酸甜、微辣，色泽黑褐，口味浓香，酸、辣、咸、甜各味谐和。在西方，辣酱油被广泛用于各种菜肴的制作中，特别是牛肉菜和牛肉制品。辣酱油也可以用于制作饮料，如血玛丽和番茄汁。

六、咖喱（curry）

咖喱是由多种香辛原料配制而成的调味品，常见于印度菜、泰国菜和日本菜等菜系。咖喱的种类很多,以国家来分,其产地有印度、斯里兰卡、泰国、新加坡、马来西亚等;以颜色来分,有棕、红、青、黄、白之别。根据配料细节上的不同来区分的咖喱有十多种。

咖喱在西餐中通常被用于烹饪肉类和瓜果,如咖喱牛肉、咖喱鸡等,但烹调时间不宜过长,否则影响菜品的口感。

第八节 西餐香料

西餐香料常分为干制香料和新鲜香草两个部分,它们在西餐烹调中能起到去除异味、增香及着色的作用。

一、干制香料

干制香料主要是将香料植物的根、茎、叶、花、果实、种子等进行干燥制作而成的,有粉末状、颗粒状和自然成型的形状等。

(一) 肉桂(cinnamon)

肉桂是肉桂树的树皮卷成条状干燥后制成。肉桂的外形有粉状、片状两种,有甜的和木头的香味。肉桂是常见的烘焙调味料,在蛋糕、曲奇和甜点制作中被广泛使用。在中东国家被运用在鸡肉和羊肉为主要食材的菜肴中。在美国,肉桂粉常和苹果搭配制作苹果派,因为肉桂可以减弱苹果的酸味。

Cinnamon is the dried bark of various laurel tree. Cinnamon sticks are made from long picccs of bark that are rolled, pressed and dried. Cinnamon has sweet, woody fragrance in both ground and stick forms, it is the most common baking spice. Cinnamon is used in cakes, cookies and desserts throughout the world, it is also used in savory chicken and lamb dishes from the Middle East. In American cooking, cinnamon is often paired with apples to make apple pie, because it can mellow the tartness of apple pie.

肉桂的香味还可以增加蔬菜和水果的口感。肉桂是巧克力的最佳搭档,可以在巧克力甜品或饮料中适量加入以增加口感和香味。粉状的肉桂不能加入沸腾的液体中,因为肉桂的香味会流失。

The flavor of cinnamon enhances the taste of vegetables and fruits. Cinnamon is a perfect partner for chocolate, we can use it in any chocolate dessert or drink. Ground cinnamon should not be added to boiling liquid, because the cinnamon will lose its flavor.

（二）香叶（bay leaf）

香叶又称桂叶，是桂树的叶子，桂树原产于地中海沿岸。香叶可分为两种：一种是月桂树的叶子，呈椭圆形，较薄，干燥后为淡绿色；另一种是细叶桂，其叶形较长且厚，背面的叶脉突出，干燥后为淡黄色。

香叶的香味清爽，微苦，干制叶、鲜叶都可以使用，用途广泛，尤其是在家庭烹饪中，在制作豆类、蔬菜汤、炖肉、意面少司等菜肴时，香叶可以起到增香的作用。香叶在使用中需要较长时间的烹煮才能有效释放其独特的香味，常用于汤汁类、肝酱类和烩肉类菜中，香叶不可以食用。

Bay leaves are pungent and have sharp and bitter taste, they are used in soups, stews, meat and vegetable dishes.

The bay leaves are useful in home-style cooking. When you are making vegetable soups, meat stews, spaghetti sauce, bay leaves can be added for more pungent flavor. Alternate whole bay leaves with meat, seafood or vegetables on skewers before cooking, be sure to remove bay leaves before eating a dish that has finished cooking, the whole leaves are used to impart flavor only.

（三）番红花（saffron）

用作香料的番红花是干燥的红花蕊雌蕊，早年经西藏走私传入我国，因此又称藏红花。在欧洲，番红花的主要产地是西班牙，因为它的栽培十分费工费时，产量又少，所以价格十分昂贵。番红花带有强烈的独特香气和苦味，在西餐中主要用于菜肴、米饭、汤及少司的调色调味，如番红花在中东地区菜肴、意大利调味饭、西班牙鸡肉饭、法式菜浓味炖鱼中被用作调色和调味的佐料。

Saffron has a spicy, pungent and bitter flavor with a sharp odor, a little pinch goes a long way with saffron. It is used in many Middle Eastern dishes, Italian risottos, Spanish chicken rice, French seafood stews and Scandinavian sweet breads.

（四）小茴香（anise seed）

小茴香为伞形科植物茴香的干燥成熟果实。在秋季果实初熟时采割植株，晒干，打下果实，除去杂质。小茴香产于地中海，质地温和，有着独特浓郁的气味，味苦且略带辛辣，有助于提神、开胃。小茴香是制作很多综合香料的主要原料，在烹饪方面，小茴香主要用于制作土豆与鸡肉等沙拉酱、肉类调味、制作烤肉酱、烹调牛肉汤等。

最好购买整个的小茴香，因为粉状的会比较容易流失香味。整个的小茴香可以保存3年。小茴香在烹煮之前，必须先将其烘烤才能将它的味道散发出来，也更利于捣碎。

将小茴香粉放入干燥洁净的瓶子里并在半年内使用完。

Anise seeds are best purchased whole, as the ground powder quickly loses its flavor. Whole anise seed keeps up to 3 years. Dry-roast anise seeds to heighten their aroma and make them brittle to crush. Store ground powder in a clean, dry jar and use within 6 months.

二、新鲜香草

新鲜香草是指香料植物在其生长过程中,采摘后直接使用的香料。

(一) 罗勒(basil)

罗勒是一种气味强劲、辛辣刺激、口感较重的香料,罗勒非常适合与番茄搭配。

Basil is a strong-smelling and strong-tasting herb that is used in cooking, it has pungent and recognizable odour. It goes well with tomatoes.

罗勒在意大利菜肴的制作中非常流行,它是pesto少司的主要原料。将捣碎的罗勒叶、大蒜、松仁、帕马森奶酪在橄榄油里混合后可搭配意面。因为罗勒很容易变色,所以在上菜前一分钟再将其加入热菜中。

Basil is very popular in Italian cooking and is the main ingredient of the popular sauce pesto, it is a sauce typically served with pasta, contains crushed basil leaves, garlic, pine nuts and Parmesan cheese in olive oil. If you add basil to warm dishes it is best to do this at the last minute, because it will discolor rapidly.

(二) 百里香(thyme)

百里香又名麝香草,其茎叶富含芳香油,主要成分有百里香酚,含量约为0.5%,百里香的叶及嫩茎可用于调味。

百里香的香气持久优雅,它带有浓郁的麝香香味,其香味经过长时间烹煮也不会消失,是熬制清汤、高汤,制作炖菜时不可缺少的香料。

Thyme has subtle, dry aroma and slightly minty flavor.

Thyme is often included in seasoning blends for poultry and stuffing and also commonly used in fish sauces, chowders, soups. It goes well with lamb and veal as well as in eggs, custards and croquettes.

(三) 迷迭香(rosemary)

迷迭香有特别清甜、带松木香的气味和风味,香味浓郁,甜中带有苦味。

在地中海地区的菜肴中经常使用,一般以较少量加入汤汁或烩菜中以获得适中的效果,主要用于羊肉、猪肉、禽类菜肴。腌制肉类时也可加入迷迭香以增添香味。

Rosemary has piney flavor, it is traditionally used in many Mediterranean dishes. Rosemary has sharp, bitter taste that will compliment a wide variety of foods such as lamb, pork, chicken and rabbit.

Rosemary's assertive flavor blends well with garlic to season lamb roasts, meat stews and marinades.

(四) 莳萝(dill)

莳萝别名土茴香、刁草,香气近似于香芹,略带清凉感,无刺激味。适用于炖菜、海鲜等菜肴。将莳萝放到汤里、生菜沙拉及一些海产品的菜肴中,有增进风味之功效。莳萝种子的香味比叶子浓郁,更适合搭配鱼虾贝类等。莳萝不能和具有刺激性气味的食物搭配,如大蒜。

Dill is a herb with yellow flowers and strong sweet smell.

Dill is a soft herb that has mild and delicate flavor, which enhances the flavor of fish, shellfish, vegetables. Dill can not combine well with pungent food like garlic.

(五) 番茜(parsley)

番茜也称荷兰芹,经常被用来装饰各种菜肴,以及制作牛油汁、混合香草、鱼汁等。其香味清烈,在西餐烹调中常用来制作法国蜗牛的香草牛油及巴黎牛油汁等,也可和其他香草混合使用,或撒在菜品的表面,以增加美感,产生香气,或当成蔬菜在汤、炖菜中使用。

Parsley is widely used in European, Middle Eastern and American cooking, it is often used as a garnish, as many dishes are served with fresh green chopped parsley sprinkled on top. Root parsley is used as a snack or vegetable in many soups, stews and casseroles.

(六) 薄荷(mint)

薄荷具有医用和食用的双重功能,其主要食用部位为茎和叶,也可将其榨汁后服用。在食用方面,薄荷既可作为调味剂,又可作为香料,还可搭配酒、茶等。薄荷叶的口感清新,在菜品制作中被广泛应用,通常被用于装饰甜品。英国、美国如果在热菜制作中使用薄荷,多搭配羊排食用。

Mint is a herb with fresh-tasting leaves.

Mint can be used in a wide range of dishes including many desserts (dessert topping) as a traditional accompaniment to lamb.

（七）阿里根奴（oregano）

阿里根奴俗称牛至，原产于地中海地区，第二次世界大战后，美国及其他美洲国家普遍种植阿里根奴。它是薄荷科芳香植物，叶子细长圆，花有一种刺鼻的芳香，在意大利菜肴中被普遍使用，是制作比萨不可缺少的香料，因此又被称为比萨草。

Oregano is an important culinary herb, used for the flavour of its leaves, which can be more flavourful when dried than fresh. It has an aromatic, warm and slightly bitter taste which can vary in intensity.

Oregano's most prominent modern use is as the staple herb of Italian cuisine, it is most frequently used with roasted, fried or grilled vegetables, meat and fish. Oregano combines well with spicy foods in southern Italy.

（八）他拉根（tarragon）

他拉根又称茵陈蒿，香味浓烈，味感与薄荷相似。常用于调制香料醋，可泡在醋内制成他拉根醋。在烹调中常与鸡肉、鱼肉、鸡蛋搭配，尤其是鸡肉，不仅可以降低鸡肉的油腻程度，还可以突出鸡肉的鲜美口感。

Tarragon is also known as estragon, it is cultivated for culinary and medicinal purposes. Fresh, lightly bruised sprigs of tarragon are steeped in vinegar to produce tarragon vinegar.

Tarragon is one of the four fines herbes of French cooking, and is particularly suitable for combining with chicken, fish and egg dishes.

（九）鼠尾草（sage）

鼠尾草又称艾草，具有强烈的香味和令人愉快的清凉感，略有苦涩味，主要用于制作鸡、鸭、猪类菜肴及肉馅类菜肴。

The pungency of sage works well to cut the fattiness of meat, so it complements chicken, duck and pork. Sage has a particular affinity to poultry, it shows up in poultry seasoning and stuffing.

（十）红椒粉（paprika）

红椒粉又称甜椒粉，红椒是茄科一年生草本植物，果实较大，呈红色，味清香，不辣，略甜，干后可制成粉，主要产于匈牙利。红椒粉在烹调中被广泛使用，常用于调色。

Paprika is used as a simple garnish for many dishes. Combine it with butter, margarine, or oil for a quick baste for fish or poultry, this is especially good on roast turkey.

（十一）香兰草(vanilla)

香兰草(vanilla)，又称香荚兰、香子兰、上树蜈蚣，多年生攀援藤木，其果实富含香兰素，香味充足。香兰草的豆荚及其衍生物在食品业中的应用十分广泛，尤其是在糖果、冰激凌及烘烤食品中。其价格仅次于藏红花蕊(saffron)，是世界上第二昂贵的调味"香料之王"。

Vanilla is one of the most popular flavorings in the world. It is used in flavoring most desserts, including ice cream, custard, cake, candy and pudding.

Vanilla beans should never be refrigerated because they may develop mold when chilled. They should be kept in an air-tight container at room temperature. Vanilla provides smooth rich background taste for shellfish, chicken and veal. Add vanilla to hot chocolate, coffee or tea for adding richness.

第九节　意大利面食

一、意大利面食的分类

意大利面食(pasta)，泛指所有源自意大利的面食，最初的记载可追溯到1154年的西西里。意大利面是用硬质小麦粉混合水、鸡蛋制成的未发酵的面团，然后制成的各种形状的面食，水煮或烤制后可食用。

Pasta is a staple food of traditional Italian cuisine, with the first reference dating to 1154 in Sicily. The noodles are made from an unleavened dough, durum wheat flour mixed with water or eggs and formed into sheets or various shapes, then cooked by boiling or baking.

意大利面分为两大类：干制的和新鲜的。大多数干制的意大利面是通过挤压制成的，新鲜的意大利面传统上都是手工制作的，但是如今各式各样的、新鲜的意大利面大都是由机器制成的。

Pasta may be divided into two broad categories, dried and fresh. Most dried pasta is commercially produced via an extrusion process although they can be produced in most homes. Fresh pasta was traditionally produced by hand, sometimes with the aid of simple machines, but today many varieties of fresh pasta are also commercially produced by large-scale machines, and the products are widely available in supermarkets.

不管是干制的还是新鲜的,意大利面都有着很多种不同的形状和种类,目前有记载的就有310种不同的形式和1300种名称的意大利面。在意大利,某种意大利面的形状或类别会经常由于地点的不同而不同。常见的意大利面有:长条形、通心粉的半月形管、包馅型等。

Both dried and fresh pastas come in a number of shapes and varieties, with 310 specific forms known variably by over 1300 names having been documented. In Italy the names of specific pasta shapes or types often vary with locale. For example, the form of cavatelli is known by 28 different names depending on region and town. Common forms of pasta include long shapes, short shapes, tubes, flat shapes and sheets, filled or stuffed and specialty or decorative shapes.

二、意大利面的煮制

煮制的意大利面需要有一些嚼劲,不能太软或成糊状。如果意大利面被过度煮制,便会失去食用它的乐趣。

Pasta should be cooked when it still feels firm to the bite, not soft and mushy. Much of the pleasure of eating pasta is its texture.

烹饪意大利面的时间根据其形状、大小、面粉的使用和含水度不同而不同。

Cooking times differ for different shape and size of pasta. Timing also depends on the kind of flour used and the moisture content. Pasta is best if cooked and served immediately.

煮制意大利面的过程如下。

Procedure for cooking pasta in large quantities.

(1) 使用4 L水、500 g的意大利面、25 g的盐。

Use at least 4 quarts boiling salted water per pound of pasta(4 L per 500 g), use about 25 g salt.

(2) 水烧开后加入意大利面,待其变软后,轻轻搅拌以免意大利面粘连。

Have the water boiling rapidly and drop in the pasta. As it softens, stir gently to prevent it from sticking together and sticking to the bottom.

(3) 继续煮开,搅拌几次。

Continue to boil, stirring a few times.

(4) 当意大利面煮好并有一定嚼劲时将其沥干并用冷水冲洗。

As soon as the pasta is al dente(咬起来硬的食物), drain it immediately in a colander and rinse with cold running water until it is completely cooled.

（5）为了避免意大利面粘连，可以加入少量的油。

Toss gently with a small amount of oil to prevent pasta from sticking.

三、意大利面酱汁

意大利面是一种简单的食物，但它的呈现形式是多样的。意大利面可以作为头盘、清淡的午餐或丰盛的晚餐，也可热食或冷食。

一份完整的意大利面制品由面条、酱汁和装饰物组成。酱料对于意大利面制品非常重要，一般情况下，意大利面酱分为红酱（tomato sauce）、青酱（pesto sauce）、白酱（cream sauce）、黑酱（squid-ink sauce）。红酱是以番茄为主制成的酱汁；青酱是以罗勒、松仁、橄榄油等制成的酱汁，其口味较为特殊与浓郁；白酱是以牛奶、奶油为主制成的酱汁，主要用于焗面、千层面及海鲜类的意大利面；黑酱是以墨鱼汁制成的酱汁，主要佐于墨鱼等海鲜意大利面。与意大利面搭配的少司汁在口味、颜色和质地上不尽相同，选择少司汁搭配意大利面时，青酱（香蒜少司）可以搭配薄的、细的意大利面，红酱（番茄少司汁）可以搭配厚的意大利面，因为厚重的少司汁能更好地附着在管状的、蝴蝶形的意大利面上。意大利面的装饰物一般根据面条形状和酱汁来选择，比较常用的有奶酪、罗勒等。

Pasta is a simple dish, but it comes in many varieties due to its versatility. Some pasta dishes are served as first course in Italy because their portion sizes are small and simple. Pasta is also prepared in light lunches, such as salads or large portion sizes for dinner. It can be prepared by hand or food processor and served hot or cold. Pasta sauces vary in taste, color and texture. When choosing which type of pasta and sauce to serve together, there is a general rule regarding compatibility. Simple sauces like pesto are ideal for long and thin strands of pasta while tomato sauce combines well with thicker pastas. Thicker and chunkier sauces have the better ability to cling onto the holes and cuts of short, tubular, twisted pastas.

（一）红酱（tomato sauce）

红酱如图 2-8（彩图 8）所示。

 原料 Ingredients

240 mL olive oil	240 mL 橄榄油
110 g onion, chopped fine	110 g 切碎的洋葱
110 g carrot, chopped fine	110 g 切碎的胡萝卜
110 g celery, chopped fine	110 g 切碎的芹菜

图 2-8

1360 g canned whole tomatoes　　　　1360 g 罐装整粒番茄
1 garlic cloves, minced　　　　　　　1 瓣剁碎的大蒜碎
15 g salt　　　　　　　　　　　　　　15 g 盐
7 g sugar　　　　　　　　　　　　　　7 g 糖

 工具 Tools

saucepan　　　　　　　　　　　　　　少司锅
wooden spatula　　　　　　　　　　　木铲

 制作方法 Method

（1）Heat the olive oil in a large saucepan. Add the onions, carrots, celery and then saute lightly for a few minutes. Do not make the vegetables brown.

先在少司锅中加热橄榄油,后加入洋葱、胡萝卜、芹菜,轻炒几分钟,不要将蔬菜炒上色。

（2）Add remaining ingredients, simmer(uncovered) for 45 minutes, until the soup is reduced and thickened.

加入剩下的原料一起炖(不加盖)45 min,直到汤汁减少变浓。

（3）Taste and adjust seasonings.

调味。

（4）For service, this sauce should be tossed with the freshly cooked spaghetti or other pasta in a bowl before being plated, rather than being simply ladled over the pasta.

装盘前直接将少司浇在意大利面上。

特点 Characters

Delicate taste, salty and sour flavor　口感细腻,浓香咸酸

在红酱的基础上可以添加其他原料制成多种少司汁搭配意大利面。

Meat sauce: Brown 500 g ground beef, ground pork, or a mixture of beef and pork, in oil or rendered pork fat, add 120 mL red wine, 1 L tomato sauce, 500 mL beef or pork stock, parsley, basil and oregano to taste. Simmer the sauce for 1 hour(uncovered).

Tomato cream sauce: Use 110 g butter instead of the olive oil in the basic recipe. At service time, add 1 cup heavy cream to the tomato sauce(250 mL per L). Bring the sauce to simmer and serve.

Tomato sauce with sausage: Slice 700 g fresh Italian sausage and brown in oil. Drain and add it to the basic tomato sauce. Simmer for 20 minutes.

（二）青酱(pesto sauce)

 原料 INGREDIENTS

100 g fresh basil leaves	100 g 新鲜罗勒叶
375 mL olive oil	375 mL 橄榄油
60 g pine nuts	60 g 松仁
6 garlic cloves	6 瓣大蒜
7 g salt	7 g 盐
150 g Parmesan cheese, grated	150 g 磨碎的帕马森奶酪
50 g Romano cheese, grated	50 g 磨碎的罗马诺干酪

 工具 Tools

blender	搅拌器
wooden spatula	木铲

 制作方法 Method

(1) Wash the basil leaves and drain well.

洗净罗勒叶后沥干水分。

(2) Put the basil leaves, olive oil, pine nuts, garlic cloves, and salt in a blender. Blend them to a paste, but not so long that the mixture is smooth. It should have a

slightly coarse texture.

将罗勒叶、橄榄油、松仁、大蒜和盐放入搅拌器里打成糊状(有粗纹理)。

(3) Transfer the mixture to a bowl and stir in the cheese.

将混合物倒入碗里拌入奶酪。

(4) To serve, cook pasta according to the basic procedure. Just before the pasta is done, stir a little of hot cooking water into the pesto to thin it, toss the drained pasta with the pesto and serve immediately.

在意大利面煮熟前,将少量热水倒入青酱中以稀释酱汁,将沥干的意大利面拌入酱汁中,最后撒上奶酪。

特点 Characters

Special and rich flavor 口味特殊、浓郁

（三）白酱(cream sauce)

原料 Ingredients

60 g butter	60 g 黄油
60 g all-purpose flour	60 g 普通面粉
500 mL milk	500 mL 牛奶
2 teaspoons of salt	2 茶匙盐
1/2 teaspoon of freshly grated nutmeg	1/2 茶匙新鲜的、磨碎的肉豆蔻

工具 Tools

saucepan	少司锅
wooden spatula	木铲

制作方法 Method

(1) In a medium saucepan, heat the butter over medium-low heat until melted. Add the flour and stir until smooth. Over medium heat, cook until the mixture turns into a light, golden sandy color, about 6 to 7 minutes.

在少司锅内小火融化黄油,待黄油融化后加入面粉,不断搅拌直至混合物变得光滑。改用中火,直至混合物变成淡黄色,需 6～7 min。

(2) Meanwhile, heat the milk in a separate pan until just about to boil. Add the hot milk to the butter mixture 1 cup at a time, whisking continuously until very smooth.

Bring to a boil. Cook for 10 minutes, stirring constantly, then remove it from heat. Season with salt and nutmeg, and set aside until it is ready to use.

与此同时,在另一个锅里加热牛奶。将牛奶分次(每次一杯)加入到黄油混合物中,不停地搅拌直至混合物变得光滑。煮大约 10 min,不断地搅拌,最后离火,加入盐和肉豆蔻调味。

特点 Characters

Creamy, smooth and delicate　　　　　　　奶香浓郁,滑爽细腻

四、意大利面菜品

(一) 奶酪通心粉(macaroni and cheese)

原料 Ingredients

450 g macaroni	450 g 意大利通心面
1 L bechamel	1 L 白少司汁
5 mL dry mustard	5 mL 干芥末
dash Tabasco	塔巴斯科辣椒酱
450 g Cheddar cheese, grated	450 g 切达奶酪,磨碎的
garnish:bread crumbs, paprika	装饰物:面包糠、红椒粉

工具 Tools

saucepan	少司锅
colander	漏勺
baking pan	烤盘
wooden spatula	木铲
Oven	烤箱

制作方法 Method

(1) Cook macaroni according to basic method for boiling pasta. Drain and rinse it in the cold water.

煮熟通心粉,沥干后,冷水过冲。

(2) Flavor the bechamel with the dry mustard and Tabasco.

加入干芥末粉和塔巴斯科辣椒酱到白少司汁中调味。

(3) Mix the macaroni with the cheese. Combine with the bechamel.

将通心粉和奶酪混合,后与白少司汁混合。

(4) Pour the mixture into a buttered pan. Sprinkle with bread crumbs and paprika.

将混合物倒入烤盘中,撒入面包糠和红椒粉。

(5) Bake at 175 ℃ until it is hot and bubbling for about 30 minutes.

放入 175 ℃的烤箱中烤大约 30 min。

(二) 千层面(lasagna)

千层面如图 2-9(彩图 9)所示。

图 2-9

 原料 Ingredients

4～5 pieces(boiled) Lasagna noodles	4～5 片千层面
20 g butter	20 g 黄油
50 g mozzarella cheese	50 g 马苏里拉奶酪
cream sauce	奶油少司
30 g butter	30 g 黄油
30 g all-purpose flour	30 g 普通面粉
250 mL milk	250 mL 牛奶
pinch salt and cinnamon	少量的盐和肉桂
tomato sauce	番茄少司
30 g butter	30 g 黄油
1 tomato	1 个番茄
100 g tomato paste	100 g 番茄酱

50 g chopped onion 50 g 切碎的洋葱

50 g sliced mushroom 50 g 切片蘑菇

50 g diced red pepper 50 g 红椒(切丁)

100 g minced meat 100 g 肉末

pinch ground black peppercorn, salt 少量盐和黑胡椒

工具 Tools

saucepan	少司锅
whisk	搅拌器
baking pan	烤盘
wooden spatula	木铲
oven	烤箱

制作方法 Method

1. Make the cream sauce

Heat 30 g butter in the saucepan over medium heat. Stir in 30 g flour. Cook and until the mixture turns into golden color. Gradually stir in 250 mL milk with a whisk. Cook and stir until the milk mixture boils and thickens. Stir in the cinnamon and salt, remove the saucepan from the heat.

在少司锅中用中火融化黄油,加入面粉搅拌直至混合物变成金黄色后,缓慢加入牛奶,不停搅拌直到混合物变稠,加入肉桂、盐,最后离火。

2. Make the tomato sauce

Heat butter in the saucepan over medium heat. Fry the onion, then add mushroom, red pepper, cook until the vegetables are tender, add in minced meat. Stir in the tomato, tomato paste, season with black peppercorn, salt, simmer for 15 minutes.

少司锅中加入黄油融化后,加入洋葱、蘑菇、红椒后炒至蔬菜变软,加入肉末。加入番茄、番茄酱搅拌,加入黑胡椒、盐调味,炖 15 min。

3. Make the layer

Baste the baking pan, then layer tomato sauce, lasagna noodles, cream sauce, and repeat for 3 times. At last add tomato sauce and cream sauce and sprinkle the cheese on the top.

在烤盘上刷油,然后依次铺上番茄少司、千层面、奶油少司,并重复3次,最后铺上番茄少司和奶油少司,在最上面撒上奶酪。

4. Bake lasagna

Loosely cover the baking pan with foil and bake for 30 minutes at 200 degrees, or until the lasagna is hot and bubbling. Set aside for 10 minutes before serving.

将烤盘用锡纸盖上,放入 200 ℃ 的烤箱中烤 30 min。烤好后静置 10 min 再端给客人。

第三章 西餐用具

第一节 厨房用具

一、西餐刀具

(一) 西餐刀(French knife or chef's knife)

西餐刀是厨房中使用最频繁的刀具,长 15～40 cm,刀头尖或圆,刀刃锋利,用途广泛,可以用来剁菜、切片、切丁等。

A French knife is the most frequently used knife in the kitchen, for general-purpose chopping, slicing, dicing and so on. Blade length of 10 inches is most popular for general work. Larger knives are for heavy cutting and chopping. Smaller blades are for more delicate work.

(二) 去皮刀(paring knife)

去皮刀为不锈钢制成,头尖刃利,长 6～10 cm,主要用于蔬菜水果的去皮、削割。

A paring knife has a small, pointed blade of 2-4 inches long, used for trimming and paring vegetables and fruits.

(三) 去骨刀(boning knife)

去骨刀长约 15 cm,刀身又薄又尖,用于肉类原料的出骨。

A boning knife has a thin, pointed blade of about 6 inches long, used for boning raw meats and poultry.

（四）切片刀（slicer）

切片刀长约 36 cm，用于切熟肉。

A slicer has a long, slender, flexible blade of about 14 inches long, used for carving and slicing cooked meats.

（五）锯齿刀（serrated knife）

锯齿刀为长形，刀刃呈锯齿状，用于切面包、蛋糕等西式点心。

A serrated knife is like a slicer, but with a serrated edge, used for cutting breads, cakes and similar items.

（六）黄油刀（butter knife）

黄油刀不需要很锋利，它主要用来抹黄油、抹果酱。

A butter knife is used for spreading butter and jam.

（七）屠刀（butcher knife）

屠刀的刀身重、刀背厚、刀刃锋利，呈弧形，用于分解大块生肉。

A butcher knife has a heavy, broad, slightly curved blade, used for cutting, sectioning and trimming raw meats.

（八）水果去皮刀（vegetable peeler）

水果去皮刀是一种小型工具，有旋转叶片，用于蔬菜、水果去皮。

A vegetable peeler is a short tool with a slotted, swiveling blade, used for peeling vegetables and fruits.

（九）磨刀（steel）

磨刀是刀具的重要组成部分，用于修整和维护刀具。

It is an essential part of the knife kit, used for truing and maintaining the knife edges.

（十）牡蛎刀（oyster knife）

牡蛎刀的刀头尖削、刀身短而薄，用于挑开牡蛎外壳。

A oyster knife is a short, rigid, blunt knife with a dull edge, used for opening oysters.

（十一）蛤蜊刀(clam knife)

蛤蜊刀的刀身短而扁平,刀口锋利,用于剖开蛤蜊壳。

A clam knife is a short, rigid, broad-bladed knife with a slight edge, used for opening clams.

（十二）砍刀(cleaver)

砍刀为方头、刃利、背厚,其刀身重,用于砍排骨等。

A cleaver has a heavy, broad blade, used for cutting through bones.

二、烹调锅具

（一）平底锅(frying pan)

平底锅又称煎锅,是一种用来煎煮食物的器具,直径为 20～30 cm,其为低锅边并且向外倾斜的铁制平底煮食用器具。现在厂商普遍改用比较轻的铝作为制造物料。平底锅适合用于煎、炒食物,容易使用,只需短短几分钟就能烹调出各式各样的佳肴。

Frying pan is used for frying, searing and browning foods. It is typically 20-300 mm in diameter with low sides that flare outwards, a long handle and no lid. The cooking surface of a frying pan is typically coated with a layer of oil or fat when the pan is used(though greasy foods like bacon do not need additional oil added).

（二）少司锅(saucepan)

少司锅是一种带有手柄的、较深的锅,用于煮制、炖和制作少司。

A saucepan is a small deep cooking pan with a handle, used for boiling, stewing and making sauces.

（三）炒锅(saute pan)

炒锅呈圆形,其为平底、形较小、较浅,锅底中央略隆起,一般用于快炒少量油脂。

A saute pan is used for general sauteing and frying meats, fish, vegetables and eggs.

（四）高汤锅(stockpot)

高汤锅又称汤桶,其桶体积大、较深,一般用于制作高汤或炖煮大量的液体。

A stockpot is a large, deep, straight-sided pot for preparing stocks and simmering large quantities of liquids.

（五）焖锅（brazier）

焖锅的锅口较宽、深度较浅，锅壁较厚，用于焖、烩菜肴等。

A brazier is a round, broad, shallow, heavy-duty pot with straight sides, used for browning, braising and stewing meats.

（六）蒸锅（double boiler）

蒸锅一般有两层，其底层盛水，上层放食物，一般用于蒸制食品。

A double boiler is a pot with two sections. The lower section holds boiling water, the upper section holds foods.

三、过滤器具

（一）漏勺（skimmer）

漏勺一般为不锈钢材质，浅底连柄，圆形广口，中间有许多小孔，用于食品油炸后沥油等。

A skimmer is used to remove(clean) the floating fat or solids from the surface of the liquid.

（二）蔬菜滤器（colander）

蔬菜滤器用于沥干洗净的蔬菜水果等。

A colander is bowl-shaped pan with many small holes in the bottom, used for separating liquid from food.

（三）筛子（sieve）

筛子用于从大块食物中分离出小块食物，从液体中分离出固体食物。

A sieve is a tool of wire or plastic net on a frame, used for separating large things from small solid pieces, or solid things from liquid.

（四）细网眼锥形滤器（chinois）

细网眼锥形滤器用于过滤少司。

A chinois is used for filtering sauce.

四、其他工具

(一) 挖球器(ball cutter)

挖球器可将蔬菜、水果等挖成球状,有不同的规格。

The blade of a ball cutter is a small, cup-shaped half-sphere, used for cutting fruits and vegetables into small balls.

(二) 厨师叉(cook's fork or chef's fork)

厨师叉是一种比较厚重的、带长柄的叉子,用于举起和将原料翻面。

A cook's fork is a heavy, two-pronged fork with a long handle, used for lifting and turning meats and other items.

(三) 直铲(straight spatula or palette knife)

直铲的刀片长且其底部是圆形的,主要用于在蛋糕上撒糖粉。

A straight spatula has a long, flexible blade with a rounded end, used mostly for spreading icing on cakes.

(四) 橡皮刮刀(rubber spatula or scraper)

橡皮刮刀是塑料材质,带有长柄,用于刮取或搅拌容器中的食物原料,混合打发的蛋清和打发的奶油。抗热的橡皮刮刀可以用于在烹饪时搅拌食物。

It is made of a broad, flexible rubber with a long handle, used to scrape bowls and pans, also used for folding in egg foam and whipped cream. Heat-resistant spatulas can be used for stirring foods while cooking.

(五) 面刀(bench scraper or dough knife)

面刀的柄是木质的,主要用于切面团。

It is made of a broad, stiff piece of metal with a wooden handle on one edge, used to cut pieces of dough.

(六) 轮刀(pastry wheel or wheel knife)

轮刀的手柄上有圆形的、转动的刀片,主用切割面团、比萨。

A pastry wheel has a round, rotating blade with a handle, used for cutting pastry and baked pizza.

（七）食品夹（food tong）

食品夹是金属制的、有弹性的"U"字形夹钳,用于夹制食物。
A food tong is a kind of tool used to pick up and handle foods.

（八）肉锤（meat tenderizer）

肉锤为铝制成,锤身有蜂窝面及平面,用于捶松及拍扁肉类,破坏肉的结缔组织,使之质嫩。
It is a hand-powered tool used to tenderize slabs of meat in preparation for cooking.

（九）擦板（grater）

擦板是一种有网格的四面金属盒子,用于将蔬菜、奶酪等食物切丝、磨碎。
A grater is a four-sided metal box with grids of varying sizes, used for shredding and grating vegetables, cheese and other foods.

（十）刮丝器（zester）

刮丝器可将橘、柠檬、橙子等的果皮刮成细丝。
A zester is a small hand tool used for removing the colored part of citrus peels in thin strips.

（十一）冰激凌球勺（ice cream scoop）

冰激凌球勺由球勺（半球形）与手柄两部分组成,勺底有一半圆形薄片,捏动手柄,细薄片可以转动,使冰激凌呈球形造型。
An ice cream scoop is a kind of spoon that can shape the ice cream.

（十二）裱花袋及花饰管（pastry bag and tubes）

花饰管也称裱花嘴,以不锈钢、铜或塑料制成。嘴部有齿形、扁形、圆口形、月牙形等各种花型。
花饰袋也称裱花袋,为布、尼龙或油纸制成的圆锥形袋子,无锥尖,在锥部开口处可插进裱花嘴,装进奶油后,可以裱花。
Pastry bag and tubes are cone-shaped cloths or plastic bags with an open end that can be fitted with metal tubes of various shapes and sizes. Used for shaping and decorating with items such as whipped cream and soft dough.

（十三）糕点刷（pastry brush）

糕点刷用于刷鸡蛋液等。

A pastry brush is used to brush items with egg wash, glaze and so on.

(十四) 开罐器(can opener)

开罐器是一种用于开罐头的装置,使用前必须清洁干净以避免污染食物。

It is a device used to open tin cans, it must be carefully cleaned and sanitized every day to prevent contamination of foods.

(十五) 打蛋器(egg whisk)

打蛋器以不锈钢丝缠绕而成,用于打发或搅拌食物原料,如鸡蛋、奶油等。

An egg whisk is used to whisk egg, cream and so on.

五、西餐计量工具

(一) 量杯(measuring cup)

量杯的材质一般为塑料、不锈钢、玻璃等,有柄,内壁有刻度,用于量取液体和固体原料。

Measuring cups are available in different sizes. They can be used for both liquid and dry measures.

(二) 量匙(measuring spoon)

量匙用于测量少量的液体或固体原料,主要用于量取调味料。

Measuring spoons are used for measuring very small volumes: 1 tablespoon, 1 teaspoon, 1/2 teaspoon, and 1/4 teaspoon. They are used most often for spices and seasonings.

(三) 长柄勺(ladle)

长柄勺用于测量液体,其大小尺度标在了勺柄上。

Ladles are used for measuring and portioning liquids. The size is stamped on the handle.

(四) 温度计(thermometer)

温度计用于测量油温、糖浆温度及肉类等的中心温度。常用的温度计种类有探针温度计和油脂、糖测量温度计,以及普通温度计等。

Thermometers measure temperature. There are many kinds for many purposes.

1. 肉类温度计(meat thermometer)

肉类温度计显示肉类的内部温度。在烹调前将其插入到食物中,烹调过程中保持温度计插在食物中的状态。

A meat thermometer indicates internal temperature of meats. It is inserted before cooking and left in the product during cooking.

2. 快速显示温度计(instant-read thermometer)

将快速显示温度计插入到食物中,几秒钟即可显示食物温度,可耐受温度可达100 ℃。当烤制食物时,不能将温度计插在食物内,以免损坏温度计。

An instant-read thermometer gives readings within a few seconds of being inserted in a food product. It reads up to 100 ℃. It must not be left in meats during roasting, or they will be damaged.

3. 油脂、糖测量温度计(fat thermometers and candy thermometers)

油脂、糖测量温度计用于测量油脂、糖浆,可耐受温度高达200 ℃。

Fat thermometers and candy thermometers test temperatures of frying fats and sugar syrups. They read up to 200 ℃.

4. 特殊温度计(special thermometers)

特殊温度计用于精确检测烤箱、冰箱等的温度。

Special thermometers are used to test the accuracy of oven, refrigerator and so on.

(五) 秤(scales)

弹簧秤(spring scale)主要用于称量各种原料。

电子秤(electronic scale)是比较精准的计量工具,能精确到小数点后一位以上。

Most recipe ingredients are measured by weight, so accurate scales are important.

第二节 西餐餐具

一、西餐餐具的发展

西餐发展的历史源远流长,据有关史料记载,早在公元前5世纪,在古希腊的西西里岛上,就出现了高度发展的烹饪文化。在当时就很讲究烹调方法,煎、炸、烤、焖、蒸、煮、炙、熏等烹调方法均已出现。在当时尽管烹饪文化有了相当的发展,但人们的用餐方法仍以抓食为主,餐桌上的餐具还不完备,餐刀、餐叉、汤匙、餐巾等都没有出现。西餐餐桌上的刀、叉、匙都是由厨房用的工具演变而来的。

西餐餐具中无论是刀、叉、汤匙还是盘子,都是手的延伸。例如盘子,它是整个手掌的扩大和延伸,而叉则更是代表了整个手上的手指。

进食用的叉子最早出现在11世纪的意大利塔斯卡地区,当时只有两个叉齿。在18世纪,由于法国的贵族偏爱用四个叉齿的叉子进餐,叉子变成了地位、奢侈、讲究的象征,随后逐渐变成必备的餐具。

二、西餐瓷器餐具

在18世纪中叶,用于西餐的瓷器餐具在欧洲普及,当时主要用作茶具和咖啡用具,中国的青花瓷传入欧洲之前,西餐中使用的用具只有金属器、玻璃器和软质陶器,中国青花瓷的精致深受欧洲人的喜爱,于是欧洲人便开始研制瓷器的,瓷器的造型、质地等不断更新,瓷器在西餐中被普遍应用。西餐瓷器餐具主要有展示盘、汤盘、汤盅、开胃品盘、面包盘、黄油盘、甜品盘、咖啡杯及垫碟、茶壶、小奶罐、糖罐等。

三、西餐玻璃餐具

玻璃餐具具有化学性质稳定、刚度高、透明光亮、清洁卫生、美观大方等优点。

西餐中常见的玻璃餐具有水杯、白葡萄酒杯、红葡萄酒杯、香槟酒杯、白兰地杯、威士忌杯、甜酒杯、直升杯、鸡尾酒杯、古典杯等,各种杯具的形状、容量、所用材料的质地有所不同。

四、西餐金属餐具

西餐金属餐具主要有银制餐具、不锈钢餐具等,在俄式服务中银制餐具被大量使用。金属餐具的具体种类和档次,要根据餐厅的档次来决定。西餐中的金属餐具主要有黄油刀、甜品刀叉勺、开胃品刀叉、主餐刀叉、鱼刀叉、汤勺、咖啡勺、蜗牛夹等,还有服务中使用的肉刀、肉叉、分餐叉勺、面包夹、冰块夹、盛汤勺、方糖夹、服务托盘、冰桶等。

第四章 西式早餐

第一节 西式早餐的分类及其特点

一、西式早餐的分类

西式早餐一般可分为三种,一种是美式早餐(American breakfast,包括英国、美国、加拿大、澳大利亚及新西兰等以英语为母语的国家的早餐都属于此类);一种是欧式早餐(Continental breakfast,包括德国、法国等国家的早餐);还有一种零点早餐。

二、美式早餐

美式早餐内容非常丰富,主要包括以下内容。

(一)果汁

果汁分为罐果汁(canned juice)和新鲜果汁(fresh juice)两种。
1. 新鲜果汁
常见的新鲜果汁包括:橙汁(orange juice),西柚汁(grapefruit juice),苹果汁(apple juice),什锦蔬菜汁(mixed vegetable juice)等。
2. 罐头果汁
常见的罐头果汁包括:蜜汁桃子(peaches in syrup),蜜汁杏子(apricots in syrup)等。

(二)谷类

玉米、燕麦等制成的谷类食品,例如玉米片(corn flakes),脆爆米(rice crispies),泡芙(puff rice)等,通常搭配砂糖、牛奶、香蕉切片或葡萄干食用。如图4-1(彩图10)所示。

麦片粥(oatmeal)、玉米粥(cornmeal)也会在早餐中提供,客人可搭配牛奶、糖调味。

图 4-1

(三) 蛋类

蛋类的营养价值丰富,是早餐的重要组成部分,根据其烹煮方法的不同,主要可以分为以下几种。

1. 煎蛋(fried eggs)

只煎一面的荷包蛋(图 4-2(彩图 11))称为 sunny-side up,两面煎半熟的(图 4-3(彩图 12))称 over easy,两面全熟的(图 4-4(彩图 13))称 over hard 或 over well-done。

图 4-2　　　　　　　　图 4-3　　　　　　　　图 4-4

2. 带壳水煮蛋(boiled eggs)

煮三分钟熟的称 soft boiled,煮五分钟熟的称 hard boiled。

3. 去壳水煮蛋(poached eggs)

将蛋去壳,滑进锅内特制的铁环中,在水面上煮至所需要的熟度。

4. 炒蛋(scrambled eggs)

煎蛋、煮蛋、炒蛋等由客人选择火腿(ham)、腌肉(bacon)、腊肠(sausage)作为配料,以盐、胡椒(peppercorn)调味。

5. 蛋卷(omelet, omelette)

蛋卷主要有以下几种形式。

普通蛋卷(plain omelet)、火腿蛋卷(ham omelet)、火腿奶酪蛋卷(ham & cheese

omelet)、奶酪蛋卷(cheese omelet)、香菇蛋卷(mushroom omelet)。

蛋卷通常用盐与辣酱(tabasco)调味,而不用胡椒,因为胡椒会使蛋卷硬化,也会留下黑斑。

(四)吐司和面包

吐司和面包通常会被烤成焦黄状,早餐还有各种糕饼,以供变换口味。注意吃的时候不可用叉子叉,要用手拿,抹上牛油、草莓酱(strawberry jam)或橘皮(marmalade),咬着吃。

在西式早餐中,常见的吐司和面包主要包含以下种类。

(1) 松饼(plain muffin):须趁热吃,从中间横切开,涂上牛油、果酱、蜂蜜或糖汁。
(2) 牛角面包(croissant):英国人则称为"crescent roll"。
(3) 压花蛋饼(waffles):可涂上牛油或枫树蜜汁食用。
(4) 法式煎蛋衣面包片(French toast):将吐司沾上蛋和牛奶调成的汁液,在平底锅中煎成两面金黄的吐司,可搭配果酱、盐及胡椒粉食用。

(五)饮料

饮料指咖啡或茶等不含酒精的饮料。西式早餐中常见的饮料主要有:咖啡、红茶、热巧克力及各种果汁等。

欧式早餐比美式早餐简单,内容大致相同,但不供应蛋类,客人需要蛋类食品时,则需要另外付费。

第二节 西式早餐制作工艺

一、西式早餐菜品制作实例

(一)蛋类

1. 炒鸡蛋(scrambled eggs)

将鸡蛋去壳打入碗中,添加适量牛奶、奶油或水,放入锅中不停搅拌至凝结,其质地柔嫩、多汁,切勿搅拌过度。用低油温将鸡蛋炒熟,保持鸡蛋的鲜嫩。

原料 Ingredients

6 eggs	6个鸡蛋
2 tablespoons of cream	2汤匙奶油
1 teaspoon of salt	1茶匙盐
1/4 teaspoon of white pepper powder	1/4茶匙白胡椒粉
3 tablespoons of salad oil or butter	3汤匙色拉油或黄油

 工具 Tools

skillet	锅
wood spoon	木勺

 制作方法 Method

(1) Whisk eggs in a small bowl until it is smooth. Mix in cream, salt and white pepper powder.

鸡蛋打入碗中,搅打至光滑,加入奶油、盐、白胡椒粉后搅拌均匀。

(2) Melt butter in a skillet over medium heat. Pour in egg mixture, cook and stir until it is set but still moist for 3 to 5 minutes.

在锅中用中火融化黄油后,倒入鸡蛋液,以木铲搅拌至凝结,需要3～5 min。

 特点 Characters

Soft texture, moist and juicy	质地柔嫩、汁多湿润

2. 煎蛋卷(omelette)

煎蛋卷(omelette,又称欧姆蛋、西式蛋饼)是将去壳打散的蛋液加入牛奶、奶油、水等,在锅中用木匙拌炒至微软定型后,再将煎好的圆形蛋饼对折成半圆形即可,蛋饼内可加入馅料。如图4-5(彩图14)所示。

 原料 Ingredients

basic omelette	基本款蛋卷
2 large eggs	2个大鸡蛋
1 small knob of unsalted butter	1小块无盐黄油
1 small handful of grated Cheddar cheese(optional)	1把切达奶酪碎

图 4-5

optional: tomato & basil omelette	番茄 & 罗勒蛋卷
2 or 3 sprigs of fresh basil	2 或 3 根新鲜的罗勒
1 handful of cherry tomatoes	1 把樱桃番茄
olive oil	橄榄油
optional: mushroom omelette	蘑菇蛋卷
2 or 3 field-type mushrooms	2 或 3 种蘑菇
olive oil	橄榄油
salt and peppercorn	盐和胡椒

 工具 Tools

frying pan	煎锅
spatula	铲子

 制作方法 Method

(1) For a basic omelette, crack the eggs into a mixing bowl with a pinch of salt and peppercorn. Beat well with a fork.

制作基本款蛋卷,在碗里打入鸡蛋后加入少量盐和胡椒,搅拌均匀。

(2) Heat a small knob of butter in a small frying pan over low heat, and once the butter is melt, add the eggs and move the pan around to spread them out evenly.

在煎锅里加入少量黄油,小火融化黄油后,加入鸡蛋液,晃动煎锅使得鸡蛋液均匀铺开。

(3) When the omelette begins to cook and firm up, but still has a little raw egg on its top, sprinkle over the cheese.

当蛋卷开始凝固,但表面上仍未熟透时,撒上奶酪。

(4) Using a spatula, ease around the edges of the omelette, then fold it over in half. When it starts to turn into golden brown underneath, remove the pan from the heat and slide the omelette onto a plate.

使用铲子松动蛋卷的边缘,然后对折蛋卷,当下面颜色变得金黄,离火后将蛋卷滑入盘中。

(5) For a tomato and basil omelette, pick and roughly tear the basil leaves. Chop the tomatoes in half, then add to a hot frying pan with a small knob of butter, a drizzle of oil and a pinch of salt and peppercorn. Fry and toss around for 1 minute, then turn the heat down to medium and sprinkle over the basil leaves. Add the eggs and move the pan around to spread them out evenly. Continue method is as same as for the basic omelette.

番茄罗勒蛋卷:撕碎罗勒叶,将番茄对半切开后,和一小块黄油、少量橄榄油、盐、胡椒一并放入锅中,煎约 1 min,然后将火调到中档,撒入罗勒叶。加入鸡蛋液并移动煎锅使得蛋液均匀铺开,做法同基本款蛋卷。

(6) For a mushroom omelette, roughly chop the mushrooms into quarters and add them to a hot frying pan with a small knob of butter, a drizzle of oil and a pinch of salt and peppercorn. Fry and toss around until it is golden, then turn the heat down to medium. Add the eggs and move the pan around to spread them out evenly. When the omelette begins to cook and firm up, but still has a little raw egg on its top, sprinkle over the Cheddar. Continue method is as same as for the basic omelette.

蘑菇蛋卷:将蘑菇一开四后,加入至放入一小块黄油的煎锅中,后加入少量橄榄油和盐、胡椒,煎至金黄色,然后将火力调至中档。加入鸡蛋液后移动煎锅使其均匀滑动。当蛋卷开始凝固,但表面上仍未熟透,撒上切达奶酪,做法同基本款蛋卷。

 特点 Characters

Soft texture, looks like crescent　　　　　质地柔嫩,形似月牙
Suggested omelet fillings: cheese, creamed mushrooms, red caviar and so on.

(二) 饼类

1. 土豆饼(hash browns)

如图 4-6(彩图 15)所示。

图 4-6

原料 Ingredients

2 potatoes	2 个土豆
2 tablespoons of chopped onion	2 汤匙洋葱碎
2 tablespoons of chopped bacon	2 汤匙培根碎
1 teaspoon of salt	1 茶匙盐
1/4 teaspoon of white pepper powder	1/4 茶匙白胡椒粉
3 tablespoons of butter	3 汤匙黄油

工具 Tools

frying pan	煎锅
wooden spatula	木铲

制作方法 Method

(1) Peel boiled potatoes, then cut them into small strips, add chopped onion, chopped bacon, salt and white pepper powder.

将土豆煮熟后去皮,切成丝状后加入碎洋葱、培根碎、盐、白胡椒粉。

(2) Add butter to the hot pan, then add potato mixture, press them to form round shape while cooking until golden brown.

待锅烧热后加入黄油,后加入土豆混合物,一边煎一边压,使其呈饼状,煎至两面金黄。

特点 Characters

Crispy outside and juicy inside, soft and crisp 外焦里嫩、松软酥香

2. 可丽饼(crepe)

可丽饼起源于法国布列塔尼省,它是一种比薄烤饼更薄的煎饼,以小麦粉制作而成的一种美食。如图4-7(彩图16)所示。

现在的布列塔尼仍保持着传统的习俗和庆典,法国人把2月2日定为"可丽饼日"。每年的2月2日庆典游行时,人们在街上、小酒馆或餐厅中手拿可丽饼,唱起歌,跳起舞,欢天喜地地庆祝丰收的来临。在布列塔尼省古色古香的小镇街道上,随处可见可丽饼专卖店。

图 4-7

 原料 Ingredients

2 eggs	2个鸡蛋
125 g flour	125 g 面粉
15 g sugar	15 g 糖
50 g butter	50 g 黄油
25 g milk	25 g 牛奶

 工具 Tools

frying pan	煎锅
wooden spatula	木铲

制作方法 Method

(1) Measure all ingredients into blender jar, blend for 30 seconds. Scrape down sides, blend for 15 seconds. Cover and let it sit for 1 hour(This helps the flour absorb more liquid).

在一个罐子里混合所有原料,搅拌 30 s,把罐子周围的混合物刮下来后,再搅拌 15 s,静置约 1 h(静置混合物可帮助面粉吸收更多的水分)。

(2) Heat frying pan with lightly grease, pour batter into pan, tilt pan to spread batter. Once crepe has lots of little bubbles, loosen any edges with wooden spatula. Flip crepe over, cook this side quickly, slide crepe from pan to plate.

加热锅后,在锅里加入少量黄油,倒入面糊,倾斜锅使得面糊铺开。一旦薄饼出现小气泡,用木铲压薄饼边缘。一面煎好后,将薄饼翻面,另一面的煎制时间会较短,最后将薄饼从锅里滑入盘中。可搭配蜂蜜、果酱食用。

特点 Characters

Golden colour, soft and tough taste　　　　色泽金黄、口感软韧

可丽饼的主要口味有以下几种。

鲜奶油可丽饼:将煎好的饼皮卷上打发好的鲜奶油一起食用,还可以切些水果碎(比如芒果,蜜桃,草莓等)和奶油一起卷。

糖奶油可丽饼:在饼皮上铺上热融的黄油,撒上白糖,就是最简单的甜式可丽饼。

糖渍苹果可丽饼:将苹果切片加糖与少许的水焖煮,然后均匀铺在饼皮上。

3. 薄煎饼(pancake)

原料 Ingredients

500 g all-purpose flour	500 g 普通面粉
60 g sugar	60 g 糖
1 teaspoon of salt	1 茶匙盐
1 tablespoon of baking powder	1 汤匙泡打粉
4 eggs	4 个鸡蛋
1 L buttermilk	1 L 脱脂牛奶
125 mL melted butter	125 mL 融化的黄油

工具 Tools

frying pan	煎锅
wooden spatula	木铲

 制作方法 Method

(1) Sift together the flour, sugar, salt, baking powder.

将面粉、糖、盐、泡打粉过筛。

(2) Mix the beaten eggs, buttermilk and butter.

混合鸡蛋、脱脂奶和融化的黄油。

(3) Add the liquid ingredients to the dry ingredients. Mix just until the dry ingredients are thoroughly moistened. Do not overmix.

将液体混合物加入面粉混合物中,不要过度搅拌混合物。

(4) Depending on the size desired, measure 30-60 mL portions of batter onto a greased, preheated griddle(190 ℃), allowing space for spreading.

根据所需要的薄饼大小,倒入 30～60 mL 的面糊到预热的、加入油的煎锅中。

(5) Griddle the pancakes until the tops are full of bubbles and begin to look dry and the bottoms are golden brown. Turn and brown the other side.

薄饼煎至底部呈金黄色后,翻面。

(6) Remove from the griddle and serve.

离火装盘。

 特点 Characters

Golden colour, soft and tough taste	色泽金黄、口感软韧

4. 华夫饼(waffle)

华夫饼也称格子饼、压花蛋饼,是一种烤饼,源于比利时,使用专用的烤盘(waffle iron)制成。如图 4-8(彩图 17)所示。

 原料 Ingredients

625 g all-purpose flour	625 g 普通面粉
1 teaspoon of salt	1 茶匙盐
2 tablespoons of baking powder	2 汤匙泡打粉
6 egg yolks, beaten	6 个蛋黄,打散

图 4-8

750 mL milk	750 mL 牛奶
250 mL melted butter	250 mL 融化的黄油
6 egg whites	6 个蛋白
60 g sugar	60 g 糖

 工具 Tools

waffle iron	烤盘
blender	搅拌机
egg whisk	打蛋器

 制作方法 Method

(1) Sift together the flour, salt and baking powder.

将面粉、盐、泡打粉过筛。

(2) Mix the egg yolks, milk and butter.

混合蛋黄、牛奶和黄油。

(3) Add the liquid ingredients to the dry ingredients. Mix just until the dry ingredients are thoroughly moistened. Do not overmix.

将液体混合物加入面粉混合物中,不要过度搅拌混合物。

(4) Whip the egg whites until they form soft peaks. Add the sugar and whip until stiff peaks form.

搅打蛋白至形成软峰后,加入糖继续搅打至蛋白发泡变硬。

(5) Fold the egg whites into the batter.

将打发的蛋白和糖的混合物加入面糊中。

(6) Pour enough batter onto a lightly greased, preheated waffle iron to almost cover the surface with a thin layer. Close the iron.

将烤盘刷油、预热后,倒入适量面糊,后盖上烤盘。

(7) Cook waffles until signal light indicates they are done or until steam is no longer emitted.

烤至成熟即可(不再有蒸汽出现)。

(8) Remove from the iron and serve.

装盘(搭配糖浆或蜂蜜食用)。

特点 Characters

Golden colour, tender and delicious　　　　色泽金黄、酥软可口

华夫饼的经典搭配如下。

华夫饼＋三明治:将各种新鲜蔬果和沙拉酱或奶油搭配在一起,夹在两片华夫饼当中。

华夫饼＋美味酱:将华夫饼涂上各种酱料、沙拉,鲜润柔滑,馥郁醇香。

华夫饼＋冰激凌:欧洲美食主义者最喜欢的吃法,冰凉爽口,浪漫怡然,风味独特。

(三) 吐司类

1. 法式吐司(French toast)(图 4-9)

图 4-9

法式吐司原本是以前的法国主妇为了不浪费食物,将隔夜变质的长棍面包、吐司面包等沾上鸡蛋液,后进行煎制或烤制,最后撒上糖粉食用。

Basic French toast consists of slices of bread dipped in a batter of eggs, milk, a little sugar and flavorings. It is cooked on a griddle like pancakes.

Bread: white pullman bread is standard. Special versions can be made with French bread, egg-rich bread or whole-grain bread.

Batter: Milk is the usual liquid, mixed with eggs in various proportions. Deluxe versions may include cream or sour cream.

Flavorings: Vanilla, cinnamon and nutmeg are popular choices.

French toast is dusted with powdered sugar, like pancakes, with accompanying butter, syrups or fruits.

The most common fault in making French toast is not soaking the bread long enough to allow the batter to penetrate. If the bread is just dipped in the batter, the final product is just dry bread with a little egg on the outside.

Cinnamon Raisin French Toast 肉桂葡萄干法式吐司

6 eggs	6 个鸡蛋
375 mL milk	375 mL 牛奶
125 mL cream	125 mL 奶油
125 g sugar	125 g 糖
1 tablespoon of vanilla extract	1 汤匙香草精华
1 teaspoon of cinnamon	1 茶匙肉桂粉
pinch of nutmeg and salt	少量肉豆蔻粉和盐
12 slices of cinnamon raisin bread(2 cm thick)	12 片肉桂葡萄干面包(2 cm 厚)
45 g butter	45 g 黄油
confectioners' sugar as needed	糖粉

saute pan	煎锅
egg whisk	打蛋器
wooden spatula	木铲

(1) Beat together the eggs, milk, cream, sugar, vanilla, cinnamon, nutmeg and salt until the sugar is dissolved.

将鸡蛋、牛奶、奶油、糖、香草、肉桂、肉豆蔻、盐一起混合,直到糖被溶解。

(2) Soak the bread in the egg mixture until the bread is soaked throughly, but do not leave it so long.

面包浸泡在鸡蛋混合物中直到面包充分被浸泡,不要浸泡太久。

(3) For each portion, heat 1.5 teaspoon of butter in a saute pan large enough to hold 2 slices of bread.

加入 1.5 茶匙黄油至煎锅中。

(4) Put 2 slices of bread in the pan and brown the bottoms.

放入 2 片面包至锅中,将其底部煎至金黄色。

(5) Turn the bread over and cook for 30 seconds.

将面包翻面后煎 30 s。

(6) Transfer the saute pan to an oven preheated 190 ℃. Bake for 10 minutes. The bread should be cooked through and slightly puffed up.

将煎锅放入预热 190 ℃ 的烤箱中烤 10 min。面包会有轻微肿胀变大。

(7) Put the bread on the plate, dust with confectioners' sugar, and serve immediately.

装盘,撒上糖粉装饰,趁热端给客人。

 特点 Characters

Golden colour, soft, tender and delicious　　色泽金黄、酥软可口

2. 三明治(sandwich)(图 4-10)

三明治是一种典型的西方食品,以面包、肉、奶酪、各种调料制作而成,吃法简便,广泛流行于西方各国。

图 4-10

公司三明治(club sandwich)是用煎蛋、火腿、蔬菜、干酪、烟肉等各式食材制作而成的。有时会制成双层形式,切成四等份,并用牙签串好,配上薯条。传统的制法是用火鸡肉、烟肉、生菜及番茄等材料,配上经烘烤的面包。

 原料 Ingredients

3 slices of white bread, toasted	3 片烤好的白吐司
mayonnaise as needed	蛋黄酱
2 leaves lettuce	2 片生菜叶
2 slices of tomato, about 0.5 cm thick	2 片番茄,约 0.5 cm 厚
3 strips of bacon	3 条培根
60 g sliced turkey or chicken breast	60 g 火鸡或鸡胸肉片

 工具 Tools

toaster	吐司机
kitchen knife	厨师刀
butter knife	黄油刀

 制作方法 Method

(1) Place the toast slices on a clean work surface. Spread the tops with mayonnaise.

将吐司放在干净的工作台上,表面涂上蛋黄酱。

(2) On the first slice, place 1 lettuce leaf, then 2 slices of tomato, then 3 strips of bacon.

在第一片吐司上放置 1 片生菜叶、2 片番茄和 3 片培根。

(3) Place the second slice of toast on top, spread the mayonnaise side down.

依次放置第二片吐司(抹蛋黄酱的一面朝下)。

(4) Spread the top with mayonnaise.

表面涂上蛋黄酱。

(5) On the top of this slice, place the turkey or chicken, then the other lettuce leaf.

吐司上放置火鸡或鸡肉、生菜叶。

(6) Place the third slice of the toast on top, spread the mayonnaise side down.

最上面放置第三片吐司(抹蛋黄酱的一面朝下)。

(7) Place frilled picks on all 4 sides of the sandwich.

在三明治的 4 个边角上插上牙签。

(8) Cut the sandwich from corner to corner into 4 triangles. Each triangle will have a pick through the center to hold it together.

将三明治切成 4 个三角形,每个三角形都有 1 根牙签固定。

(9) Place the sandwiches on a plate with the points up. The center of the plate maybe filled with potato chips, French fries, or other garnish or accompaniment.

装盘,将三明治和炸薯条或其他装饰物一起搭配摆放。

 特点 Characters

Crispy outside and soft inside　　　　　　松脆软嫩

第五章　西餐少司的制作工艺

高质量的少司可以给菜肴增添口味,增加用餐者的食欲,厨师需要掌握制作中的技巧,才能领悟烹制的精髓。

第一节　少司的概念与作用

一、少司的概念

少司是英文 sauce 的音译,还可译为沙司或调味汁。它是用于确定菜品滋味的黏稠润滑的液体,是由厨师调制西餐菜肴和糕点的调味汁。少司具有确定菜肴味道,增加菜肴的营养价值,增加菜肴的美观性等作用。许多原料在烹调过程中都会产生一些汁液,这是原料的原汁,不能当成少司。少司的制作技术是西餐制作中最重要的技术之一,在西餐厨房中,制作少司被列为一项单独的工作,少司由有一定经验的厨师专门制作。这种少司与菜肴主料分开制作的方法,是西餐烹调的一大特点。

二、少司的作用

(一) 味道(flavor)

在少司制作中加入各种原料和调味制品,使其具有风味,可以给予或增加菜品的味道。因为少司大多有一定的浓稠度,所以当少司淋在菜肴上时,可以起到保温的作用。有一定浓稠度的少司隔开了原料与空气,减缓了原料中热量的散失。对菜品热量的保存也是对菜品风味的保存,这使得菜肴更新鲜,味道更鲜美。

（二）浓郁（richness）

在不影响原料自身味道的基础上，不同的少司制作用料能给食物带来不同的特色风味，从而形成不同特色的经典菜式。

（三）湿润（moistness）

在菜肴上淋上少司，如同菜肴外部形成了保护膜，能减少菜肴水分的流失，避免原料的表面形成干皮，增加其滋润度，从而提升菜肴风味。

（四）外观色泽与亮度（appearance）

人们在面对色彩鲜艳的菜肴时，会表现出更强的食欲。西餐菜肴讲究可食性，在菜盘中的所有食材，必须可以食用，而且西餐厨师以客人吃完菜肴、盘中不剩余一点食材为荣耀。因此，西餐厨师在装盘菜肴或糕点时，常常将各种色泽鲜艳的少司作为装饰料，这样既可以引导客人去尝试食用菜盘中的所有食材，又能起到独特的美化效果。

（五）开胃（appetizing）

少司作为菜肴的重要组成部分，可以丰富菜肴的味道，增强人们的食欲。

第二节　少司的组成

一、液体（liquid）

大多数少司的主体或基底为液体食材。大多数都是使用以下 5 种不同的液体食材调制而成。由此调制而成的 5 种基本少司，称为主少司或母少司。

（1）用纯牛奶制成贝夏梅尔少司（Bechamel sauce）。

（2）用白色基础汤（white stock）制成瓦鲁迪少司（Veloute sauce）。

（3）用棕色基础汤（brown stock）制成布朗少司（Brown sauce）或西班牙少司（Espagnole sauce）。

（4）用番茄加基础汤制成番茄少司（tomato sauce）。

（5）用澄清的黄油（clarified butter）制成荷兰少司（Hollandaise sauce）。

二、稠化剂(thickening agent)

少司的稠度必须达到一定的程度,才能黏附于食物上。稠度不够的少司,则会似水般在餐盘中肆意流动,但也不能过于浓稠。调制少司时,淀粉类是最为普遍使用的稠化剂。面粉是在淀粉类中最常用于稠化少司的食材之一。其他可用于稠化的淀粉质类有:玉米淀粉、葛粉、即溶淀粉、胶化淀粉、面包粉,除淀粉质类以外还有其他蔬菜与谷物类制品,如洋芋淀粉、米面粉等。

(一)淀粉质

1. 面油是由面粉与黄油按1:1的比例调制成的光滑无颗粒的泥糊状物质。在少司即将烹调完成前加入面油,可以达到使少司迅速稠化的目的。当其中的黄油溶化后,不但可以增添少司的味道,还可以增加食物的可观性,让人食欲大增。使用面油时,仅需将少量的面油放入加热的少司中,用打蛋器搅散均匀至光滑状。后重复相同的步骤,直到调制成所需的浓稠度。为了将面粉的生涩味去除,少司离火前,应将其额外加热数分钟。

2. 用玉米淀粉调制而成的少司,质地较清,颜色光亮。使用时,仅需要将玉米淀粉与冷的液体搅拌至光滑状,再放入加热的液体中。煮沸后,改以小火慢煮至液体清澈且不含淀粉质生涩的味道即可。稠化后的少司不宜煮沸过久,否则淀粉质会分解并失去稠化能力,使少司变稀。玉米淀粉的稠化能力几乎是面粉的2倍。甜点、甜点少司,以及某些特定肉类的佐餐甜味少司,皆使用玉米淀粉作为稠化剂。

3. 葛粉的用途与玉米淀粉相同。由葛粉调制而成的少司相较于玉米淀粉更为清澈。但成本较高,故极少使用。

4. 黍腊粉适用于需要被冷冻的少司。黍腊粉不同于面粉以及其他淀粉质,它不会受冷冻的影响,它的使用方法与玉米淀粉相同。

5. 胶化淀粉或即溶淀粉都是经过事先烹煮过或胶化后干制而成的。因此,它们不需要加热即可将冷的液体食材稠化,这种淀粉较少用于调制少司。

6. 面包粉或其他碎粉类,与即溶淀粉相似,因为它们皆经烹调过,其稠化速度比较快。面包粉仅适用于质地与光滑度不需考量的少司。

(二)油糊的制备

油糊必须经一定时间的烹煮以去除淀粉生涩的口感。依据烹饪加热的不同分为以下三类。

1. 白油糊(white roux) 加热时间较短,去除面粉的生涩味即可;或加热至气泡呈现,质地似砂粒般,色泽仍为白色即可。白油糊适用于调制贝夏梅尔少司,以及其他加入牛奶制成的白少司。

2. 棕油糊(blond roux or pale roux) 加热时间稍长,至油糊的颜色开始变深时即可。棕油糊用于调制 veloutes 少司,以及其他使用白高汤制成的少司。由棕油糊制成的少司,皆呈现象牙白的色泽。

3. 褐油糊(brown roux) 将油糊加热至呈现淡褐色,并冒出坚果般气味时即可。加热时,必须以小火进行,从而将油糊完全褐化而不至于烧焦。面粉在加入油脂之前放入烤箱中烘烤褐化可调制成深褐色的油糊。深褐色的油糊,虽然只有白油糊稠化能力的 1/3,但却带给少司不同的风味与深褐的色泽。

(三) 蛋黄与乳脂稠化剂

蛋黄因其蛋白质受热后会变性凝结,故具有将少司稠化的能力。当使用蛋黄作为稠化剂时,应特别注意加热的温度。若加热温度过高,蛋黄中的蛋白质将会凝结成块,并与汁液分离而形成团块颗粒。

蛋黄凝结的温度为 60～70 ℃。为了避免蛋黄在烹饪中过早凝结,将蛋黄与适量的高脂鲜奶油调和,可将其凝结温度提高至 82～85 ℃。其可承受的最高温度仍低于沸点,烹煮时仍应留意。高脂奶油除可提高蛋黄受热的温度,还可增加少司的稠度与味道。

事实上,蛋黄的稠化能力比较弱,使用它的主要目的是增添少司浓郁的风味,以及使其质地柔顺,稠化为附带效果。因为蛋黄不稳定,所以仅用于少司的最终调和步骤。

(四) 调味与调香原料

虽然少司的主体及味道是由高汤烹煮调制而成,但是仍应加入其他的材料,将少司的基本风味予以适当的变化甚至修饰至完美的境界。传统的西餐少司,根据加入基本少司中的特定调香材料而予以分门别类。

第三节 少司的基本分类

西餐中几乎所有的少司都是从以下五个主要少司衍生出来的,而它们一般很少被直接使用,需要加入其他食材加工而成为各种不同的少司。

(1) 白色少司:milk ＋ white roux ＝ Bechamel sauce(贝夏梅尔少司)

(2) 白色基础汤少司:white stock ＋ white or blond roux ＝ Veloute sauce(瓦鲁迪少司)

(3) 棕色基础汤少司:brown stock ＋ brown roux ＝ Brown sauce or Espagnole(布

朗少司或西班牙少司）

(4) 红色少司：tomato ＋ roux ＝ tomato sauce（番茄少司）

(5) 黄油少司：butter ＋ egg yolks ＝ Hollandaise sauce（荷兰少司）

第四节　少司制作实例

一、白色少司

现介绍贝夏梅尔少司（Bechamel sauce）的制作实例。

 原料　Ingredients

50 g butter	50 g 黄油
60 g flour	60 g 面粉
1.3 L milk	1.3 L 牛奶
salt, peppercorn	盐、胡椒
nutmeg	肉豆蔻

 工具　Tools

saucepan	少司锅
whisk	打蛋器
spoon	炒勺
fine chinois	细漏勺

 制作方法　Method

1. Melt butter in a saucepan. Add flour, stir and saute without coloring.
在少司锅中融化黄油，后加入面粉，搅拌、翻炒，不要使面粉上色。

2. Add cold milk and stir till sauce is boiling with a whisk. Reduce heat, season with salt, peppercorn and nutmeg.
加入冷牛奶，搅拌直至其沸腾后，减小火力，加入盐、胡椒和肉豆蔻调味。

3. Simmer for 15-20 minutes. Strain through a fine chinois.

煮 15～20 min，用细孔漏勺过滤即可。

二、白色基础汤少司

（一）黏稠少司（鸡肉少司）(Velouté sauce(chicken sauce))

 原料　Ingredients

180 g butter	180 g 黄油
200 g flour	200 g 面粉
1.5 L chicken stock	1.5 L 基础鸡汤
salt，peppercorn	盐、胡椒

 工具　Tools

saucepan	少司锅
whisk	打蛋器
spoon	炒勺
fine chinois	细漏勺

 制作方法　Method

1. Melt the butter in the saucepan. Add flour, stir and saute without coloring.
用黄油炒香面粉，不能使面粉炒上色。
2. Add the cold chicken stock, simmer for 25-30 minutes, stir occasionally.
加入冷的基础鸡汤，小火炖 25～30 min，偶尔搅拌。
3. If necessary add more stock and season. Strain through a fine chinois.
必要时添加更多的基础汤，调味后过滤。

（二）苏伯汉姆少司(sauce of chicken "Supreme")

 原料　Ingredients

700 mL chicken sauce	700 mL 鸡肉少司
150 mL cream	150 mL 奶油
18 g butter	18 g 黄油
20 mL lemon juice	20 mL 柠檬汁

salt, peppercorn　　　　　　　　　　　　　　盐、胡椒

工具 Tools

saucepan	少司锅
whisk	打蛋器
spoon	炒勺
fine chinois	细漏勺

制作方法 Method

1. Stir the sauce to desired consistency.
搅拌少司调至所需的稠度。
2. Add cream and butter.
加上奶油和黄油。
3. Add lemon juice to taste and season.
加入柠檬汁后调味。

三、棕色基础汤少司

(一) 牛骨烧汁(beef demi-glace)

demi-glace,这种用牛骨熬制的浓汁,几乎成为半数传统少司的鼻祖。demi-glace 最早是由法国大厨奥古斯特-埃科菲(Auguste Escoffier)创作的,它由烤过的牛骨加上小牛肉汤及一些配料制作而成。做好这道酱汁的最佳方法:在烤箱中每隔 20 min 把牛骨翻一次面,并连续烤制 1 h 以上,然后再放入汤中用小火熬制 20 h 以上。其目的是充分挥发深入骨髓的香气,再将牛骨中所有的胶原蛋白融入汤中,这样做出的 demi-glace 有着浓厚的牛肉味,通常将其作为其他酱汁的基础汁。

原料 Ingredients

2 tablespoons of tomato paste	2 汤匙番茄酱
1.2 kg beef bones	1.2 kg 牛骨
1.8 L cold water	1.8 L 冷水
300 g chopped onion	300 g 切碎的洋葱
150 g chopped celery	150 g 切好的芹菜
150 g chopped carrot	150 g 切好的胡萝卜

6 g black peppercorn	6 g 黑胡椒
flat-leaf parsley sprigs	法香
thyme sprigs	百里香
bay leaves	香叶

 工具 Tools

stockpot	汤锅
fine chinois	细漏勺
cook knife	西餐刀
baking sheet	烤盘

 制作方法 Method

1. Preheat oven to 220 ℃. Roast bones about 60 minutes in the preheated oven until well-browned.

预热烤箱至 220 ℃,烤牛骨大约 60 min 直至其上色。

2. Drizzle oil onto a baking sheet while bones are roasting, spread onion, carrots, and celery onto baking sheet, spread tomato paste over the vegetables and mix to roast vegetable mixture in the preheated oven, bake at 200 ℃ for 40 minutes or until well-browned.

烤骨头时,把油滴到烤盘上,将洋葱、胡萝卜和芹菜铺在烤盘上,将蔬菜和番茄酱混合后放入预热后的烤箱,200 ℃烤 40 min 直至褐色。

3. Place bones in a large stockpot. Add cold water carefully to roasting pan, scraping pan to loosen browned bits. Pour liquid from pan into the stockpot, then bring to a boil. Reduce heat, and simmer for 20 hours, skim surface occasionally, discarding foam.

把骨头放在大汤锅里,往烤盘里加入冷水,刮烤盘使里面棕色的部分变松。把液体从烤盘中倒入汤锅,煮至沸腾后减小火力,炖 20 h,撇去表面浮沫。

4. Strain broth through a fine chinois into a large container set in an ice bath, chill to room temperature.

用细漏勺过滤后,冷却。

5. Cover the container with a lid or plastic wrap and chill in the refrigerator until demi-glace is cold and set.

盖上盖子或保鲜膜,放入冰箱冷藏,直至冷却并凝固。

(二) 黑胡椒汁 (black peppercorn sauce)

 原料 Ingredients

1.2 L red wine sauce	1.2 L 红酒汁
45 g butter	45 g 黄油
65 g shallots, chopped	65 g 青葱,切碎的
16 g black peppercorn	16 g 黑胡椒
100 mL red wine	100 mL 红葡萄酒
20 mL brandy	20 mL 白兰地酒

 工具 Tools

saucepan	少司锅
whisk	打蛋器
spoon	炒勺
cook knife	西餐刀

 制作方法 Method

1. Saute black peppercorn and shallots, add to brandy and red wine, reduce by 50%.

将黑胡椒、青葱炒香,加入白兰地、红葡萄酒,浓缩至 50%。

2. Add sauce, reduce by 1/3, season to taste.

加入少司汁,当锅中汤汁减少 1/3 后,调味。

(三) 牛骨髓汁 (bordelaise sauce)

 原料 Ingredients

60 g shallots, peeled	60 g 青葱,去皮
2 g white peppercorn	2 g 白胡椒
120 mL red wine	120 mL 红葡萄酒
70 mL demi-glace	70 mL 烧汁
280 g bone marrow	280 g 牛骨髓
3 g thyme, fresh	3 g 鲜百里香
bay leaves	香叶适量

20 mL lemon juice　　　　　　　　　　　20 mL 柠檬汁
salt, peppercorn　　　　　　　　　　　　盐、胡椒

 工具　Tools

saucepan　　　　　　　　　　　　　　少司锅
spoon　　　　　　　　　　　　　　　　炒勺
fine chinois　　　　　　　　　　　　　　细漏勺
cook knife　　　　　　　　　　　　　　西餐刀

 制作方法　Method

1. Place red wine, chopped shallots and crushed peppercorn in the saucepan.

将红葡萄酒、切碎的青葱和压碎的胡椒放入少司锅中。

2. Bring to boil and reduce to a quarter, add demi-glace, simmer for 10-20 minutes.

将锅中液体烧开,当量减少至四分之一时,添加烧汁,煮 10~20 min。

3. Adjust seasoning, strain through a fine chinois.

调味后过滤。

4. Cut the bone marrow into either cubes or slices and blanch them in the water. Then add lemon juice.

将骨髓切成小块或片,焯水后放入少司中,并加入柠檬汁。

(四) 红酒汁(red wine sauce)

 原料　Ingredients

60 g shallots, peeled and chopped　　　60 g 青葱,去皮、切碎
200 mL red wine　　　　　　　　　　　200 mL 红葡萄酒
9 g rosemary　　　　　　　　　　　　　9 g 迷迭香
800 mL demi-glace　　　　　　　　　　800 mL 烧汁
salt, peppercorn　　　　　　　　　　　盐、胡椒
55 g butter　　　　　　　　　　　　　　55 g 黄油

工具 Tools

saucepan	少司锅
spoon	炒勺
cook knife	西餐刀
fine chinois	细漏勺

 制作方法 Method

1. Place red wine, chopped shallots, rosemary and crushed peppercorn into saucepan.

在少司锅中放入红葡萄酒、切碎的青葱、迷迭香和碎胡椒。

2. Bring the mixture to boil and reduce to a quarter. Add demi-glace, simmer for 10-15 minutes, adjust seasoning.

将混合物煮沸后使其量减少到四分之一，加入烧汁煮 10～15 min 后，调味。

3. Strain through the fine chinois, whisk in butter flakes to thicken.

过滤后，放入黄油搅拌，使其浓度变稠。

（五）波特酒汁（port wine sauce）

波特酒属于强化葡萄酒，其口感通常是丰富的、甜美的、厚重的。波特酒汁可用来搭配各种肉类菜肴。

 原料 Ingredients

100 mL red wine	100 mL 红葡萄酒
12 g shallots, peeled and chopped	12 g 青葱，去皮、切碎
2 g white peppercorn	2 g 白胡椒
600 mL demi-glace	600 mL 烧汁
100 mL port wine	100 mL 波特酒
25 g butter	25 g 黄油
3 g salt	3 g 盐

 工具 Tools

saucepan	少司锅
fine chinois	细漏勺

制作方法 Method

1. Place red wine, shallots and white peppercorn in a saucepan, bring to boil and reduce until the wine has evaporated.

将红酒、青葱和白胡椒放在少司锅里,待其烧开后用小火炖煮直至酒水浓缩。

2. Add demi-glace, then simmer for 8-12 minutes, add port wine, strain through a fine chinois.

放入烧汁,用小火加热8~12 min后,倒入波特酒,过滤。

3. Gradually mix in butter flakes, adjust seasoning.

逐渐加入黄油后,调味。

四、红色少司

番茄少司(tomato sauce)制作实例如下。

 原料 Ingredients

100 g onion, chopped	100 g 切碎的洋葱
10 g garlic, chopped	10 g 切碎的大蒜
30 g oil	35 g 油
100 mL water	100 mL 水
1 kg tomatoes	1 kg 番茄
55 g tomato paste	55 g 番茄酱
8 g salt	8 g 盐
12 g sugar	12 g 糖
3 g basil	3 g 罗勒
1 g white pepper powder	1 g 白胡椒粉

 工具 Tools

saucepan	少司锅
spoon	炒勺
cook knife	西餐刀

制作方法 Method

1. Saute garlic, onions in oil until fragrant.
将大蒜、洋葱爆炒至有香味传出。

2. Add tomato paste and fry for 5 minutes. Add tomato and water, bring to boil.
加入番茄酱,炒 5 min 后,放入番茄、水,煮沸。

3. Season to taste.
调味。

五、黄油少司

荷兰汁(Hollandaise sauce)制作实例如下。

 原料 Ingredients

55 g shallots, peeled and chopped	55 g 青葱,去皮、切碎
200 mL white wine	200 mL 白葡萄酒
2 g white peppercorn	2 g 白胡椒
7 egg yolks	7 个蛋黄
300 g butter	300 g 黄油
4 g salt	4 g 盐
2 mL tabasco	2 mL 塔巴斯科辣椒酱
20 mL lemon juice	20 mL 柠檬汁

 工具 Tools

saucepan	少司锅
spoon	炒勺
fine chinois	细漏勺
bain marie	双层锅(图 5-1(彩图 18))

 制作方法 Method

1. Place shallots, wine and white peppercorn in a small saucepan.
在一个小的少司锅里放入青葱、白葡萄酒和白胡椒。

2. Bring to boil and reduce to 150 mL, strain through a fine chinois into a mixing bow.

将混合物煮开并浓缩至 150 mL,过滤到碗里。

3. Add egg yolks to the mixture to form a thick cream consistency.

加入蛋黄至混合物中,使其形成浓缩奶油状。

4. Over low heat, mix in clarified butter very slowly, beat constantly, season with salt, tabasco and lemon juice.

小火搅拌,慢慢加入澄清的黄油,将其搅拌均匀后,用盐、辣椒酱和柠檬汁调味。

5. Keep at about 50 ℃ in a bain marie.

放入温度约为 50 ℃ 的双层锅里保温。

图 5-1

第六章　西式沙拉的制作工艺

第一节　沙拉的分类及其特点

一、沙拉的定义

沙拉是指将加工处理后的原料与调味品混合在一起而成的各式各样的菜品。绿色沙拉一般是以生菜或其他绿叶蔬菜为基料,然后再加上其他配料,如切碎的蔬菜、水果、坚果、奶酪、肉类或其他可食性原料,再在上面淋上沙拉汁。用料不同的沙拉在餐桌上的用途也不相同,一般可作为头盘、主菜、配菜、甜品等。将沙拉作为一餐的开始是最基本的类型,午餐期间的汤和沙拉也是很受欢迎的组合。沙拉可以单独食用,也可以和肉、鱼或鸡肉搭配食用。蔬菜沙拉、肉类沙拉和意大利面沙拉都是常见并受大众喜爱的沙拉种类。

Salad refers to a variety of dishes with foods processed of raw materials and condiments mixed up. A green salad generally based on lettuce or some other leafy green vegetables, then topped or tossed with other ingredients such as chopped vegetables, fruits, nuts, cheeses, meats, or any other edible ingredients. Different ingredients in salads make them play different roles in menu, which can be used as appetizer, main course, side dish, dessert and so on. Salad is usually served as a appetizer, and soup with salad also be a popular combination during lunch time. Salad could be only one dish in a meal or combined with meat, fish or chicken dish. Vegetable salads, meat salads and pasta salads are all common and popular salad types.

二、沙拉的分类

(一) 以原料与沙拉汁配合的形式分类

Simple salad consists of two types of ingredients and served with a dressing.

普通沙拉是由两种原料组成并另外配上沙拉汁。

Mixed salad consists of more than one type of ingredient, usually mixed together with a suitable dressing.

混合沙拉是由一种以上的原料组成的,通常混合搭配合适的调味料。

(二) 以功能分类

头盆沙拉(appetizer salad)、佐餐沙拉(accompaniment salad)、主菜沙拉(main-course salad)、独菜沙拉(separate-course salad)、甜品沙拉(dessert salad)。

三、沙拉的基本组成

(1) 基本原料:大部分沙拉都使用沙拉菜(叶用蔬菜)。

Salad greens are mostly used in salad making.

(2) 主要原料:指在沙拉中作为主要部分或沙拉强调突出的部分。

The ingredient is the main part of a salad or a salad highlighted.

(3) 沙拉汁(酱):用来调味和润滑原料,给沙拉增添更多风味。

Dressing is used to season and lubricate the ingredients, add more flavor to the salad.

(4) 装饰物被用来增添沙拉的颜色、风味并美化其外观。

Garnishment is used to add color, flavor for salad, it must complement the art of one dish.

四、沙拉酱汁的作用

(1) 增加沙拉的风味。

Add flavour and taste to the salad.

(2) 有助于原料间的搭配。

Give better combination between ingredients.

(3) 增加沙拉的营养价值。

Increase the nutritional content of salad.

（4）有利于消化。

Good for the digestion when eating salad.

五、沙拉酱的分类

（一）制作沙拉酱的原料

制作沙拉酱的原料种类繁多，大部分的沙拉酱都是以下面三种调味汁为基础制作而成的。

（1）油醋汁(vinaigrette dressing)。

（2）蛋黄酱(mayonnaise)。

（3）其他乳化调味汁(emulsified vinaigrette dressing)。

（二）制作要点

（1）必须使用高质量的原材料。

We must use good quality ingredients.

（2）如果有新鲜的香料，要用新鲜的香料代替干制香料。

If fresh herbs are available, use them instead of dried herbs.

（3）避免使用有强烈味道和气味的油。

Avoid using oil that has a strong flavour.

（4）当制作低脂肪或低胆固醇的沙拉汁（酱）时，一定要使用低胆固醇或低脂肪的油。

When making low fat or low cholesterol dressing, we must use low fat or low cholesterol oil.

（5）使用防凝固的油制作沙拉汁（酱），因为做好的沙拉汁要放进冰箱。

Use anticoagulant oil for making dressing, because salad must be refrigerated.

（6）使用油或剩余沙拉汁之前应确保它们无腐臭的味道，扔掉任何变质的原材料。

Make sure oils or leftover dressings are not rancid before using them. Throw away any rancid ingredients.

（7）制作沙拉汁时应使用生鸡蛋，不用将鸡蛋煮熟，例如制作蛋黄酱和恺撒汁。

Eggs for making salad dressings should not be cooked, such as mayonnaise and Caesar salad dressing.

（8）如果沙拉汁需要添加柠檬汁，应采用新鲜的柠檬榨汁并将其核和果肉过滤。

If the salad dressings need to add lemon juice, we should squeeze fresh lemon and strain out seeds and flesh.

第二节 沙拉的制作工艺

一、沙拉酱的制作实例

(一) 蛋黄酱(mayonnaise)

 原料 Ingredients

10 egg yolks	10 个蛋黄
25 g mustard	25 g 芥末
100 mL white wine vinegar	100 mL 白酒醋
13 g salt	13 g 盐
2.5 L corn oil	2.5 L 玉米油
12 mL lemon juice	12 mL 柠檬汁
2 g white peppercorn	2 g 白胡椒

 **工具 Tools**

bowl	碗
egg whisk	打蛋器

 **制作方法 Method**

(1) Place egg yolks, mustard, white wine vinegar, salt and white peppercorn in a mixing bowl.

将蛋黄、芥末、白酒醋、盐、白胡椒放入碗中混合。

(2) Mix all ingredients then gradually stir in the corn oil.

将所有成分混合后,逐渐放入玉米油搅拌。

(3) Add lemon juice and season.

加入柠檬汁并调味。

特点 Characters

Light and creamy　　　　　　　　　　细腻滑软

（二）千岛酱（thousand island dressing）

千岛酱是一种受欢迎的沙拉酱,适用于搭配蔬菜、三明治、生菜和海鲜沙拉等。

Thousand island dressing is a popular mayonnaise based on salad dressings and a delicious dipping. Serve it with vegetables, sandwiches, lettuce and seafood salads.

 原料 Ingredients

200 g pickled cucumber, chopped fine	200 g 酸黄瓜碎
110 g onions, chopped fine	110 g 洋葱碎
110 g bell peppercorn	110 g 灯笼椒
10 g parsley, chopped	10 g 法香碎
15 g paprika	15 g 甜红椒粉
10 mL tabasco	10 mL 塔巴斯科辣椒酱
120 g ketchup	120 g 番茄酱
1 kg mayonnaise	1 kg 蛋黄酱

 工具 Tools

knife	西餐刀
bowl	碗
egg whisk	打蛋器

 制作方法 Method

(1) Combine all ingredients in a small bowl. Stir well.

将所有的原料放入一个小碗里,搅拌均匀。

(2) Place dressing in a covered container and refrigerate for several hours, stir occasionally.

把调好的汁放在带盖的容器里冷藏几小时,偶尔搅拌。

特点 Characters

Light and creamy   细腻滑软

（三）油醋汁(vinaigrette dressing)

原料 Ingredients

200 mL red wine vinegar   200 mL 红酒醋
50 g mustard   50 g 芥末
20 g garlic   20 g 大蒜
3 g paprika   3 g 甜红椒粉
3 g black peppercorn   3 g 黑胡椒
20 g shallots   20 g 青葱
10 mL tabasco   10 mL 塔巴斯科辣椒酱
20 mL worcestershire sauce   20 mL 伍斯特少司
8 g salt   8 g 盐
600 mL olive oil   600 mL 橄榄油

 工具 Tools

bowl   碗
egg whisk   打蛋器

 制作方法 Method

(1) Roast shallots and garlic, then chop them finely.
将青葱和大蒜烘烤后切碎。
(2) Whisk together all ingredients except the oil.
搅拌除油以外的所有成分。
(3) Add oil.
加入油混合。
(4) Keep in cool place, stir well before serving.
放至凉处，使用前搅拌均匀。

（四）恺撒汁(Caesar dressing)

恺撒汁在 1924 年由厨师恺撒·卡帝尼(Caesar Cardini)发明于墨西哥一个小镇,起初名为 Aviator's 沙拉,后来才为人们所熟悉并以 Caesar(恺撒)为名。

The original Caesar salad(dressing) was made in Tijuana, Mexico by Caesar Cardini in 1924, it was first known as Aviator's salad.

 原料 Ingredients

4 egg yolks	4 个鸡蛋黄
50 mL lemon juice	50 mL 柠檬汁
400 mL olive oil	400 mL 橄榄油
50 g anchovies	50 g 银鱼柳
15 g garlic, peeled	15 g 大蒜,去皮
1 g white pepper powder	1 g 白胡椒粉
20 g mustard, Dijon	20 g 法式芥末
5 g salt	5 g 盐
50 g Parmesan cheese	50 g 帕玛森奶酪

 工具 Tools

cook knife	西餐刀
bowl	碗
egg whisk	打蛋器

 制作方法 Method

(1) Finely chop garlic and anchovies.

将大蒜和银鱼柳切碎。

(2) Place egg yolks, mustard, vinegar and lemon juice in mixing bowl.

把蛋黄、芥末、醋和柠檬汁混合。

(3) Add oil slowly while whisking.

逐步分次加入油。

(4) Add remaining ingredients, adjust seasoning.

加入剩下的原料混合均匀后,调味。

(五)泰式牛肉沙拉调味汁(Thai beef salad dressing)

 原料 Ingredients

50 g garlic, peeled	50 g 大蒜,去皮

120 g coriander, fresh	120 g 新鲜的香菜
1 g white pepper powder	1 g 白胡椒粉
50 g sugar	50 g 糖
100 g Thai fish sauce	100 g 鱼露
100 mL lime juice	100 mL 青柠汁

 工具 Tools

cook knife	西餐刀
bowl	碗
blender	搅拌器

 制作方法 Method

Mix all ingredients in the blender to marinate Thai beef salad.

将所有的原料放入搅拌器内进行混合,用于泰式牛肉沙拉调味。

二、沙拉的制作实例

(一) Mixed salad with balsamic vinegar dressing, goat cheese and croutons

混合沙拉配黑醋酱、山羊奶酪和面包丁(图 6-1(彩图 19))

 原料 Ingredients

30 g lettuce	30 g 生菜
1 lemon	1 个柠檬
10 g shallot	10 g 青葱
40 g goat cheese	40 g 山羊奶酪
30 g croutons	30 g 面包丁
olive oil	橄榄油
balsamic vinegar	黑醋
salt and black peppercorn	盐和黑胡椒

 工具 Tools

knife	西餐刀
salad bowl	沙拉碗

salad spoon   沙拉匙

 制作方法 Method

(1) Mix shallot, vinegar and olive oil in a bowl, add salt, black peppercorn and lemon.

将青葱、醋和橄榄油放入沙拉碗中混合,并加入盐、黑胡椒和柠檬。

(2) Clean all the salad. Mix the goat cheese with black peppercorn.

把沙拉全洗干净,把山羊奶酪和黑胡椒混合在一起。

(3) Serve with hot croutons.

配上热面包丁。

图 6-1

(二) 尼斯沙拉(nicoise salad)(图 6-2(彩图 20))

 原料 Ingredients

300 g tuna fish	300 g 金枪鱼
300 g potatoes, cooked and sliced	300 g 马铃薯,煮熟、切片
25 g red kidney beans	25 g 红芸豆
35 g onions, sliced	35 g 洋葱,切片
10 g black olives	10 g 黑橄榄
5 pieces of anchovies	5 片银鱼柳
200 g tomatoes	200 g 番茄
150 g leaf salad	150 g 叶状沙拉
vinaigrette dressing	油醋汁

工具 Tools

knife	西餐刀
salad bowl	沙拉碗
salad spoon	沙拉匙

 制作方法 Method

(1) Toss ingredients gently with vinaigrette dressing.
将油醋汁放入原料中拌匀。
(2) Garnish with black olives, anchovies and tomato wedges.
放入黑橄榄、银鱼柳、番茄作装饰。

图 6-2

第七章　西式汤菜的制作工艺

第一节　基　础　汤

基础汤也称原汤(stock)或汤底,是制作少司、汤、焖烩菜肴等的基本液体原料,通常以牛肉、鸡肉、鱼肉以及它们的骨头作为主要原料,配以蔬菜、调味料和水长时间炖制而成。法国烹调大师曾说:"基础汤在烹调过程中意味着一切,没有它将一事无成。"

一、原料的选择与加工

(一)原料的选择

1. 骨头

骨头是基础汤最重要的原料,它决定了基础汤的基本香气、色泽和味道,常用的骨头包括牛骨、鸡骨和鱼骨等。因为不同种类的原料需要炖制的时间也不同,所以在制作基础汤时一般不混合使用不同种类的动物原料。

(1)牛骨最好选用新鲜的、年龄较小的牛骨。小牛骨比成年牛骨含更多软骨和结缔组织,其胶原蛋白含量较高,在炖制过程中,胶原蛋白会转化为胶质和水,从而增加高汤的浓度和香味。牛骨最好选用背骨、颈骨和腿骨,因为这些部位的胶原蛋白含量更高。

(2)鱼骨最好选用含脂肪少的鱼类(如龙利鱼、鲆鱼、鳕鱼等)的骨头,脂肪含量高的鱼类的骨头中脂肪含量也较高,会影响汤的颜色、味道和状态。

(3)鸡骨(火鸡骨)最好选用鸡的颈部和背部的骨头,或者是整只鸡的骨架。

(4)其他骨头。羊骨、猪骨、野味骨等也可以使用,但不宜混用。

2. 蔬菜香料

蔬菜香料是指洋葱、胡萝卜和芹菜这三种蔬菜以2∶1∶1的比例混合而成的混合物。胡萝卜、芹菜无须去皮,用于基础汤的制作,可增进滋味和香味,通常被切成块段使用,形状的大小根据炖汤的时间而定,时间越短,形状越小。

3. 调味料

基础汤的主要调味料有:胡椒籽、香叶、百里香、番茜枝等,一般制成用纱布袋包装的"香草束"使用。标准香料包由胡椒籽、香叶、蕃茜枝、百里香组成。

(二) 原料的加工

牛骨应加工成 8 cm 左右长的段,以便在炖制过程中更充分地释放其滋味,更有利于营养物质溶于水中。

整鸡的骨架应剁成小段。

整条鱼骨应剁成小件,鱼头应一劈为二,再剁小件。

所有骨头在炖煮之前都应用冷水将血污等冲洗干净,以免影响汤的滋味及色泽。

二、基础汤的种类

传统的西餐基础汤按照使用原料和成色的不同可细分为很多类型。现代的西餐基础汤类型被简化区分,按色泽基本可以分为以下两大类。

(一) 白色基础汤

白色基础汤是将各种肉骨主料、调味蔬菜和香料放入冷水中,用小火慢煮而成的高汤。因汤的色泽清亮,近似无色,故称白色基础汤。白色基础汤根据使用的动物原料的不同,可分为白色牛骨基础汤、白色牛肉基础汤、白色鸡肉基础汤、白色猪肉基础汤、蔬菜基础汤等。

(二) 褐色基础汤

褐色基础汤又称布朗基础汤,是指将各种肉骨主料、调味蔬菜和香料送入烤炉烤至棕褐色,或在燃气灶上煎制成棕褐色之后加水用小火慢煮而成的棕褐色高汤。

通常根据肉骨主料的差异可细分为:布朗小牛基础汤、布朗鸡肉基础汤、布朗鸭肉基础汤、布朗鸽肉基础汤、布朗羊肉基础汤、布朗野味基础汤等。

三、基础汤的制作工艺流程与要求

(一) 牛肉基础汤(白色及褐色)

1. 白色牛肉基础汤
1) 制作方法

以牛肉或骨头为主料熬制,配以蔬菜和调味香料。制作白色基础汤通常要先用冷水清洗动物原料,完全去除异味和血水后,冷水下锅,再用大火煮到微沸,把汤面上的浮沫去除,随后转至小火炖制1~2 h即成。关火后要过滤汤水,撇去汤表层的浮油后即可使用,或待其冷却后密封冷藏保存。

2) 注意事项

(1) 选料要选择新鲜且无腐败异味的原料。

(2) 将蔬菜切成小丁,牛肉或牛骨切后的大小形状应尽量保持一致。差异过大会导致生熟度难以把控。

(3) 汤中浮沫一定要去除干净,不然会影响汤的成色和味道。

(4) 汤中不能使用盐来调味,否则会影响汤汁的原味与营养。

2. 褐色牛肉基础汤

1) 制作方法　其原料与白色牛肉基础汤的原料无太大的区别,只是工艺手法有些不同,并加入了番茄或番茄酱。

(1) 将牛肉、牛骨加工成小块,蔬菜切成小丁,准备好香料包。

(2) 将烤箱预热至240 ℃,将牛肉、牛骨放入烤盘烤30~40 min,使其均匀上色,中途可翻动。去除烤盘上的油,然后放入蔬菜,继续烤至蔬菜上色。

(3) 将烤制好的蔬菜倒入锅中,加入番茄酱炒匀,待蔬菜料炒至深棕色时取出备用。

(4) 在烤盘上加入冷水,后加热,将盘底的牛油和原料煮稠,取出备用。

(5) 将牛骨汤煮5 h后加入炒香的蔬菜调料,还有煮稠的汁水,继续小火煮至汤汁醇浓,大约耗时1 h,汤色呈褐色时即成。过滤,撇去浮游杂物,备用或密封冷藏。

2) 注意事项

(1) 选料要选择新鲜且无腐败异味的原料。

(2) 用高温炒制时要注意翻面,避免烤焦牛肉、牛骨。

(3) 可炒制或高温烤制蔬菜,有利于上色、增加香味,含水分较多的蔬菜不宜烤制,可炒香上色。

(4) 牛骨汤中的浮沫油要不断去除以保证汤的清澈。

(二) 鱼基础汤

1. 制作方法

(1) 采用海鲜鱼骨熬煮鱼清汤,将蔬菜切小丁,香料放入纱布香料包内备用。

(2) 将蔬菜用热油炒香,后加入鱼骨一起翻炒均匀,倒入白葡萄酒煮稠,加入冷水、香料包,大火煮至微沸后,调成小火不加盖慢煮35~45 min,中间需不断去除浮沫、浮油和杂质。

(3) 将煮好的鱼汤过滤,撇去浮油即可备用,或冷却后密封保存。

2. 注意事项

(1) 要选用脂肪少、腥味少的鱼骨、鱼肉来制作鱼清汤。如龙利鱼、大比目鱼、牙鳕鱼、无须鳕等,这样做出的汤汁是清澈不黏稠的,味鲜香。

（2）加入冷水至水刚好淹没鱼骨,煮制时间不宜过长,不然鱼骨的腥味会被煮出来。

（三）鸡基础汤

1. 制作方法

（1）将鸡的内脏和血污去除干净,然后去除鸡骨,将骨头切成 8～10 cm 的大块,蔬菜切大块,制作香料包备用。

（2）将鸡骨用冷水洗净,去除其异味和血水后,加入锅中,倒入冷水至淹没鸡骨和鸡肉,用大火煮制至微沸,去除浮沫后转小火继续煮制,保持微沸状态。

（3）煮制 2 h 后,加入蔬菜和香料袋,用小火慢煮 1～2 h 即成。期间不断去除浮沫、浮油。将汤过滤后,去除杂质即可备用,或密封冷藏。

2. 注意事项

（1）若是冷冻过的鸡肉、鸡骨应在焯水后把水倒去,煮汤需另加冷水煮制。

（2）煮汤时不可以用大火,要保留汤的鲜美。

（3）鸡肉、鸡骨的油质较多,要注意随时撇去,保证汤的清澈。

四、基础汤的制作要点

（1）加冷水下锅,使血污得以快速凝固并浮于水面,方便去除。

（2）采用温火炖煮,可以使汤汁中的血污和脂肪慢慢煮出,便于去除,也能更好地保留原汁原味。

（3）在小火慢煮期间应不断去除杂质,保持汤的清澈。将汤煮好后过滤,使其口感更好。

（4）需冷藏的基础汤,应使其尽快冷却,以免滋生细菌。

五、基础汤的质量鉴别

基础汤成品的质量通常根据风味、色泽和清澈度来鉴别,以成品的汤色纯正、汤面无浮油和浮沫、汤汁清澈、无食物残渣、香味浓郁为标准。

（1）无论哪一类基础汤,都要根据配方和标准程序制作,从而使汤的整体风味平衡,汤质醇浓,主料香味突出,蔬菜味清香,香料味清淡适宜。

（2）汤无油腻感。

（3）汤中无咸味,只有原汤的鲜味和蔬菜的原味。

（4）白色基础汤通常汤汁清澈,加热时色泽呈淡金黄色。

（5）褐色基础汤通常汤汁呈深琥珀色或棕褐色。

（6）蔬菜基础汤的色泽通常因蔬菜种类的不同而各有差异。

第二节　西餐汤的种类

西餐汤一般以基础汤或水为基本原料，与不同的配料和调味料搭配熬煮而成。在西方人的饮食中，吃主食之前常利用汤菜来开胃润喉，为进餐做好准备，因此这类汤常常被称为开胃汤。

一、汤的分类

汤的种类比较多，制作方法也多，通常情况下分为以下几种。

（一）清汤

清汤，法国称为 consomme 或 bouillon，英国称为 broth，通常是在西餐高汤的基础上，经过调味、添加或不添加配料制作而成的。清汤有各种味道，可分为牛肉清汤、鸡肉清汤、鱼肉清汤等。清汤的制作利用了蛋白质热变性的原理。为了让蛋白质溶于水中，先把瘦肉、蛋清等加入水搅匀放置一小时。当把瘦肉、蛋清等加入基础汤后，用木铲搅动，可使蛋白质和汤液充分接触。这样可使得加热后的蛋白质变性凝固，同时，也使汤液中的其他悬浮物质凝固在一起，通过过滤，使汤液变得更加清澈。

（二）浓汤

浓汤是指加入了增稠原料使之具有一定浓稠度的汤，分为奶油汤（cream soup）、菜蓉汤（puree soup）、虾贝浓汤（bisque）和杂烩浓汤（美国称为 chowder，法国称为 potage）。制作浓汤时通常会加入奶油、番茄酱等用来增加汤的浓度。由于配料不同，各种浓汤的味道也不同。例如：菜蓉汤是将蔬菜、豆类用清水煮烂、磨碎过滤后加入清汤或忌廉汤及烟肉皮用慢火熬成的，食用时用炸面包丁（croutons）或打成泡沫的忌廉在汤上作点缀，如甘笋茸汤、菠菜茸汤等。忌廉汤是以油性炒面粉加入牛奶、清汤、忌廉和一些调味品制成的汤类，以此为基汤加入鱼、鸡肉、蔬菜泥等可制成品种不同的忌廉汤，如蘑菇忌廉汤、蔬菜忌廉汤等。

（三）蔬菜汤

蔬菜汤是以油和蔬菜作为原料，然后加清汤调制的汤类。此类汤大多含有肉类，又被称为肉类蔬菜汤，其中可分为以下几种：以牛肉清汤为基汤制成的牛肉蔬菜汤，如洋葱汤、牛尾汤；以鸡清汤为基汤制成的鸡蔬菜汤，如鸡杂浓汤；以海鲜清汤为基汤制成的汤，如龙虾汤。

（四）冷汤

冷汤即冷食的汤，是用清汤或凉水加上蔬菜和少量肉类调制而成。根据是否加热可划分为热制冷食汤和冷制冷食汤。

二、好汤的必备条件

（1）使用较好的基础汤与原料。
（2）使用适当的烹调与储存方法。
（3）使用适当的服务方式。

三、汤的配料与装饰

西餐汤风味别致，花色多样，在世界各地不同的地区都有不同的展现。例如法国的洋葱汤、意大利的蔬菜汤、俄罗斯的罗宋汤等。汤除了主料以外，常常还会在汤的面上放一些小料加以补充和装饰，常用的有以下几种。

（1）炸面包丁：将面包切成丁放入黄油中炸或炒成金黄色。
（2）菜丁：将块茎类蔬菜切成丁。
（3）奶酪：把奶酪切成小片或碎末。
（4）菜丝：将蔬菜切成很细的丝。
（5）无味的饼干：如苏打饼等。
（6）荷兰芹（番茄碎）。
（7）培根切片、炒香等都可作为汤的配料使用，以增加汤的整体效果。

以上这些配料，往往起到画龙点睛的作用，可达到意想不到的效果。不同类型的汤，其装饰与配料的搭配也不同。主体与客体相平衡，互相润色，呈现的效果才是和谐和完整的。

第三节 汤的制作实例

一、基础汤的制作

（一）白色高汤（basic white stock）

 原料 Ingredients

3 kg bones(chicken, veal, beef)	3 kg 骨头（鸡骨、小牛骨、牛骨）
6 L water	6 L 水
mirepoix	植物性调味料
160 g onion	160 g 洋葱
80 g celery	80 g 芹菜
80 g carrot	80 g 胡萝卜
sachet	香料包
dried bay leaf	干香叶
1 teaspoon of dried thyme	1 茶匙干制百里香
1/3 teaspoon of peppercorn	1/3 茶匙胡椒
parsley stems	法香
whole cloves	丁香粒

 工具 Tools

stock pot	汤锅
fine chinois	细漏勺

 制作方法 Method

(1) Cut the ingredient into 8-10 cm pieces and clean them with cold water.
将原料切割成 8～10 cm 的块状，用冷水清洗去除杂质。

(2) Place them in a stockpot, cover them with cold water then boil. Skim the

surface.

将切好的原料放入汤桶中,加入冷水浸没原料,大火烧开后转小火,撇去汤表面的浮沫。

(3) Add the mirepoix and sachet ingredients.

加入植物性调味料和香料袋。

(4) Simmer for required time, skim the surface as often as necessary.

根据要求时间煮制,经常除去浮沫。

注意事项如下:

(1) 牛、小牛:6~8 h。

Beef and veal:6-8 hours.

鸡:3~4 h。

Chicken:3-4 hours.

(2) 加水量必须保持浸过骨头。

Add enough water to cover bones.

(3) 在冷水中快速冷却汤体,冷藏。

Cool the stock in a cold water bath and refrigerate.

(二) 褐色高汤(basic brown stock)

 原料 Ingredients

3 kg bones(veal or beef)	3 kg 骨头(小牛骨或牛骨)
6 L water	6 L 水
mirepoix	植物性调味料
220 g onion	220 g 洋葱
110 g celery	110 g 芹菜
110 g carrot	110 g 胡萝卜
100 g tomato puree	100 g 番茄酱
sachet	香料包
1 dried bay leaf	一片干香叶
dried thyme	干制百里香
peppercorn	胡椒
parsley stems	法香
whole cloves	丁香粒

工具 Tools

stock pot　　　　　　　　　　　　　　　汤锅
fine chinois　　　　　　　　　　　　　　细漏勺

 制作方法 Method

(1) If bones are whole, cut them into about 8 cm long pieces with a meat saw. Heat the oven to 200 ℃, place bones in the oven and brown them well.

如果是整块骨头,可将其用骨锯切割成 8 cm 左右的块,后放入 200 ℃ 的烤箱,烤制至褐色。

(2) Remove bones from pan to stockpot. Cover them with water and bring to a simmer. Skim and let the stock continue to simmer.

从烤盘中取出骨头并放至汤桶中,用冷水浸没骨头,持续小火煮制,并去除表面的浮沫。

(3) Brown vegetables well in the oven. Add the browned mirepoix, the tomato product, and the sachet to the stockpot.

将蔬菜放入烤箱烤至上色后,与番茄酱和香料袋一同放入汤桶内。

(4) Continue to simmer for 6-8 hours, skim the surface as necessary, then strain.

继续煮制 6~8 h,期间不时去除浮沫,最后过滤。

(5) Cool the stock in a cold water bath, and refrigerate.

在冷水中快速冷却汤体后,冷藏保存。

(三) 鱼基础汤(fish stock)

 原料 Ingredients

3 kg bones from lean fish　　　　　　　　3 kg 鱼骨
60 g mushroom　　　　　　　　　　　　60 g 蘑菇
30 g butter　　　　　　　　　　　　　　30 g 黄油
mirepoix　　　　　　　　　　　　　　　植物性调味料
　120 g onion　　　　　　　　　　　　　　120 g 洋葱
　60 g celery　　　　　　　　　　　　　　60 g 芹菜
　60 g carrot　　　　　　　　　　　　　　60 g 胡萝卜
　250 mL white wine(dry)　　　　　　　　250 mL 干白葡萄酒
sachet　　　　　　　　　　　　　　　　香料包

1/2 dried bay leaf　　　　　　　　　　半片干制香叶
1/3 teaspoon of peppercorn　　　　　1/3 茶匙胡椒
parsley stems　　　　　　　　　　　　法香
whole cloves　　　　　　　　　　　　 丁香粒
4 L cold water　　　　　　　　　　　　4 L 冷水

 工具 Tools

stock pot　　　　　　　　　　　　　　汤锅
fine chinois　　　　　　　　　　　　　细漏勺

 制作方法 Method

(1) Put the butter on the bottom of a heavy stock pot. Place the mirepoix in the bottom of the pot and the bones over the top of it.

将黄油放在汤锅底部,后将植物性调味料和鱼骨放入汤锅。

(2) Set the pot over low heat and cook slowly for about 5 minutes, until the bones are opaque and begin to exude juices.

用小火加热 5 min,直到骨头微微变色,并开始渗出汤汁。

(3) Add the wine, bring to a simmer, then add the sachet and water to cover them, bring to a simmer again, skim, and let them simmer for 35-45 minutes.

加入干白葡萄酒,小火煮制,然后加入香料袋和水,小火慢煮 35～45 min,撇去汤表面的浮沫。

(4) Strain the stock.

过滤。

(5) Cool stock in a cold water bath, and refrigerate.

在冷水中快速将汤体冷却,冷藏保存。

(四) 蔬菜基础汤(vegetable stock)

 原料 Ingredients

45 mL olive oil　　　　　　　　　　　45 mL 橄榄油
mirepoix　　　　　　　　　　　　　　植物性调味料
　　400 g onion　　　　　　　　　　　　400 g 洋葱
　　200 g celery　　　　　　　　　　　　200 g 芹菜
　　200 g carrot　　　　　　　　　　　　200 g 胡萝卜

200 g leek	200 g 大葱
125 g mushroom, sliced	125 g 蘑菇,切片
15 g garlic, chopped	15 g 大蒜,切碎
200 mL white wine(dry)	200 mL 干白葡萄酒
sachet	香料包
1/2 dried bay leaf	半片干制香叶
peppercorn	胡椒
parsley stems	法香
whole cloves	丁香粒
9 L cold water	9 L 冷水

 工具 Tools

stock pot	汤锅
fine chinois	细漏勺

 制作方法 Method

Sweat the onion, garlic, leek in 45 mL olive oil before adding the remaining ingredients. Cook the stock for 30-45 minutes.

用橄榄油将洋葱、大蒜、大葱炒香出汁,后放入其他原料,煮制 30～45 min 即可。

二、汤类菜品的制作

(一)蘑菇汤(mushroom soup)

 原料 Ingredients

240 g mushroom, diced	240 g 白蘑菇,切丁
1 onion	1 个洋葱
35 g butter	35 g 黄油
2 tablespoons of oil	2 汤匙油
600 mL chicken stock	600 mL 鸡基础汤
2 tablespoons of flour	2 汤匙面粉
160 mL milk	160 mL 牛奶
salt, pepper	盐、胡椒

工具 Tools

stock pot	汤锅
fine chinois	细漏勺
cook knife	西餐刀

 制作方法 Method

(1) Clean and halve mushrooms.

把蘑菇清洗干净后切成两半。

(2) Fry onion in the butter for 2 minutes.

加入黄油炒香洋葱 2 min。

(3) Add mushrooms for a further 4 minutes.

加入蘑菇后再炒 4 min。

(4) Sprinkle flour into the pan and stir well to remove lumps, add half of the stock.

将面粉撒入平底锅中,搅拌均匀,加入一半的基础汤。

(5) Stir and scrape cooked flour off the bottom, add half of the milk, stir all the time.

搅拌并刮下底部的面粉,加入一半的牛奶,搅拌。

(6) Continue to add stock and milk, adjust seasoning.

继续加入基础汤和牛奶,再放入胡椒和盐调味。

(7) Simmer very gently for 20 minutes and strain the soup.

小火慢煮 20 min 后过滤。

(二) 菠菜汤(spinach soup)(图 7-1(彩图 21))

 原料 Ingredients

300 g fresh spinach	300 g 新鲜的菠菜
3 potatoes	3 个土豆
vegetable stock	蔬菜汤
2 tablespoons of butter	2 汤匙黄油
peppercorn, salt	胡椒、盐
cream	奶油

工具 Tools

stock pot　　　　　　　　　　　汤锅
blender　　　　　　　　　　　　搅拌器
cook knife　　　　　　　　　　　西餐刀

 制作方法 Method

(1) Wash and clean the fresh spinach.
将新鲜的菠菜洗净。
(2) Peel the potatoes and cut them into big chunks.
将土豆清洗去皮后，切成大块。
(3) Add the butter to melt in a pot, then add the chopped potatoes and spinach. Cook for 5 minutes, until the spinach gives its juice.
锅中放入黄油，然后加入土豆和菠菜。炒 5 min 直至菠菜出汁。
(4) Add the vegetable stock.
加入蔬菜汤。
(5) Bring to a boil then reduce to simmer for 20 to 25 minutes, or until the potatoes are cooked. Use blender to mix the soup.
煮至沸腾后，小火煮 20~25 min，直到土豆煮熟，再用搅拌器混合。
(6) Add a dash of cream.
加入少量奶油即可。

图 7-1

(三) 南瓜汤 (pumpkin soup)

 原料 Ingredients

1 tablespoon of olive oil	1 汤匙橄榄油
1 small onion	1 个小洋葱
2 cloves of garlic	2 瓣大蒜
2 teaspoons of curry powder	2 茶匙咖喱粉
salt and peppercorn to taste	盐、胡椒适量
400 mL chicken broth	400 mL 鸡基础汤
200 mL milk	200 mL 牛奶
300 g pumpkin puree	300 g 南瓜泥

 工具 Tools

stock pot	汤锅
blender	搅拌器

 制作方法 Method

(1) Heat olive oil in a large pot.

倒入橄榄油至锅中。

(2) Saute the onion and garlic.

炒香洋葱、大蒜。

(3) Add the curry powder, salt and peppercorn to the pot.

加入咖喱粉、盐、胡椒。

(4) Add the chicken broth and pumpkin to the pot and simmer for about 15 minutes.

加入鸡基础汤、南瓜泥到锅中炖 15 min。

(5) Put batches in blender to make puree.

分批倒入搅拌器中打成泥状。

(6) Pour back into the pot and add the milk.

倒入锅中,最后加入牛奶。

（四）烟熏三文鱼杂烩浓汤（smoked salmon chowder）

 原料 Ingredients

a knob of butter	一小块黄油
1 onion, diced	一个洋葱，切丁
750 g potatoes, diced	750 g 土豆，切丁
500 mL chicken stock	500 mL 鸡基础汤
500 mL milk	500 mL 牛奶
350 g smoked salmon, cut into ribbons	350 g 烟熏三文鱼，切成条状
a handful of parsley leaves, chopped	一把切碎的欧芹
1 lemon, halved	一个柠檬，对半切

 工具 Tools

stock pot　　　　　　　　　　　　　　汤锅

 制作方法 Method

(1) Fry the onion gently in the butter, then add the potatoes, chicken stock, milk and simmer until the potatoes are very tender.

加入黄油炒香洋葱，后加入土豆、鸡基础汤、牛奶，炖至土豆变软。

(2) Add the smoked salmon and parsley leaves and season well.

加入烟熏三文鱼和欧芹后，调味。

(3) Heat everything through and add a squeeze of lemon.

将混合物煮透后，最后挤上柠檬汁。

第八章　西餐热菜的制作工艺

第一节　以油传热的烹调方法

以油为传热介质是一种常见的烹调方法,多数油脂经加热后温度可高达200 ℃以上。用油来烹制食物可使菜品成熟快,并具有良好的风味和脂香气,但也会破坏食物中的营养素。以油传热的烹调方法主要有炸、煎、炒等。

一、炸(deep frying)

(一)概念

炸是指把加工成型的原料经过调味、挂糊或不挂糊后放入炸锅中,用旺火多油(浸没原料)加热至成熟并上色的烹调方法。

Deep frying means to cook food submerged in hot fat until cooked and has golden colour. The most often used equipment for this cooking method is the deep-fryer, many foods are given a coating of breading or batter before being deep-fried.

(二)质量标准及特点

成品呈金黄色,外焦里嫩,水分流失率少,脂香气浓郁,无油脂异味,风味良好。
Standards of quality for deep-fried foods.
(1) Minimal fat absorption.
(2) Minimal moisture loss.
(3) Attractive golden color.
(4) Crisp surface or coating.
(5) No off-flavors imparted by the frying fat.

（三）分类

1. 清炸

将原料改刀成型，经过调味后直接放入油中炸制的方法。如炸薯条、炸排骨等。

2. 干炸

将原料改刀成型，调味后，蘸上一层干面粉或淀粉直接入油炸制。如炸鲜虾。干面粉对油脂有较大的破坏作用。

3. 挂糊炸（batter）

将原料改刀成型，调味后，在表面裹上面糊，入油炸制。

4. 面包糠炸（breading）

将原料改刀成型，调味后拍面粉、蘸鸡蛋、挂面包糠，后入油炸制。

（四）面糊的种类

炸制蔬菜时，软的、易成熟的蔬菜可以直接炸制，其他较难成熟的蔬菜需要预先烹制（蒸或煮），然后在其表面裹上面糊或面包糠糊，这样可以减少炸制时间。

Tender, quick cooking vegetables can be fried raw, others may be precooked by simmering or steaming briefly then coated with breading or batter to reduce the cooking time in the frying fat.

Raw vegetables are always breaded or dipped in batter before frying, such as eggplant, onion ring, tomato.

Blanched or precooked vegetables are always breaded or dipped in batter before frying, such as carrot, cauliflower.

禽类和鱼类在炸制前也经常需要挂糊，面糊的质量直接影响成品的质量。

Poultry and fish items are always breaded or dipped in batter before frying, the quality of the breading or batter affects the quality of the finished product.

食物挂糊后炸制有以下优势。

Most foods to be deep-fried are first given a protective coating of breading or batter, this coating serves three purposes.

更好地保持食物的水分和风味。

阻止食物吸收更多的油脂。

赋予食物酥脆和诱人的外观。

Help to retain moisture and flavor in the product.

Protect the food from absorbing too much fat.

Give crispness and nice appearance to the food.

（1）英式面糊：面粉 30 g，面包粉 10 g，鸡蛋 60 g。

(2) 法式面糊:面粉 50 g,牛奶 50 mL。

(3) 酵母面糊:面粉 200 g,牛奶 250 mL,酵母 10 g,盐 5 g。

(4) 蛋白面糊:面粉 200 g,牛奶 250 mL,蛋清 2 个,盐 5 g。

(5) 啤酒椰奶面糊:面粉 200 g,啤酒 250 mL,椰奶 25 mL,盐 5 g。

(6) 泡打粉面糊:面粉 200 g,牛奶 250 mL,泡打粉 5 g,盐 5 g。

(7) 蛋黄糊:面粉 100 g,蛋黄 3 个,盐 2 g。

(8) 鸡蛋糊:先拍面粉,再蘸鸡蛋液入油炸制。

(五) 适用范围

炸制食物需要使食物在短时间内成熟,因此适合制作粗纤维少、水分充足、易成熟的原料。如鱼虾类、嫩肉等。

(六) 制作步骤

Basic procedure for deep-frying.

(1) Collect all equipments and foods. 准备原料和用具。

(2) Heat the fat to proper temperature. 加热油到合适的温度。

(3) Season the food and coat it with the desired coating. 食物调味后挂糊。

(4) Add the food to the hot fat. 将食物放入油锅。

(5) Turn the food items over when cooking so that all sides of the food brown evenly. 将食物翻面(为了能上色均匀)。

(6) Remove the food from the fat and place it on the clean paper to drain. 食物起锅后放在吸油纸上以吸掉多余的油脂。

(7) Serve immediately while the food is hot. 食物趁热端给客人。

(七) 操作要点

(1) Fry at proper temperature.

Most foods are fried at 175-190 ℃. Excessive greasiness in fried foods is usually caused by frying at too low temperature.

大多数食物的油炸温度为 175~190 ℃,低温油炸会使食物变得油腻。

(2) Overload the baskets will lower the fat temperature.

装满油炸筐会降低油脂的温度。

(3) Use good quality of fat, the best fat for frying has a high smoke point.

使用优质的油脂(发烟点较高的)。

(4) Replace the fat with fresh fat after each use.

每次炸完食物后,更换油脂。

(5) Avoid frying strong and mild flavored foods in the same fat.

避免把味道重的食物和味道淡的食物放在同一批油脂中炸制,以免串味。

(6) Fry as close to service as possible.

炸制好后第一时间起锅。

(7) Remove excess moisture from foods before frying. Dry baskets.

炸制前去除食物多余的水分并确保油炸筐是干燥的。

(8) Rinse baskets and kettles well after frying.

炸完后,清洁油炸筐,洗净后消毒。

(八) 制作实例

1. 油炸洋葱圈(onion rings)

200 g flour, sifted	200 g 筛过的面粉
1 egg, beaten	1 个鸡蛋,打散
1/2 teaspoon of salt	1/2 茶匙盐
1/4 teaspoon of white pepper powder	1/4 茶匙白胡椒粉
1 teaspoon of baking powder	1 茶匙泡打粉
15 g oil	15 g 油
250 mL water	250 mL 水
250 g onion	250 g 洋葱

cook knife	西餐刀
mixing bowl	拌菜碗
deep-frier	深油炸锅
skimmer	漏勺

(1) Mix flour, salt, white pepper powder, baking powder in a bowl.

在碗里混合面粉、盐、白胡椒粉、泡打粉。

(2) Add the dry ingredients to the liquid to fold into smooth batter.

将干的原料放入液体原料混合物中,折叠形成光滑的面糊。

(3) Peel the onion and cut crosswise into 0.5 cm slices. Separate into rings.

洋葱去皮后切成厚度为 0.5 cm 的圈状。

(4) Place the onionrings in cold water, if they are not used immediately, to maintain their crispness.

若不立即使用,洋葱圈可浸泡在冷水中以保持其脆度。

(5) Drain and dry the onionrings thoroughly.

沥干洋葱圈。

(6) Dredge with flour and shake off excess.

洋葱圈拍粉。

(7) Dip a few pieces at a time in the batter and fry in deep fat(175 ℃)until golden brown.

洋葱圈挂面糊后放入 175 ℃的油温中炸至金黄色即可。

(8) Drain and serve immediately.

沥油,端盘服务。

| Soft inside and crisp outside | 外酥内软 |

2. 油炸鸡排(deep-fried chicken)(图 8-1(彩图 22))

600 g chicken	600 g 鸡肉
125 g flour	125 g 面粉
1 teaspoon of salt	1 茶匙盐
1/2 teaspoon of white pepper powder	1/2 茶匙白胡椒粉
3 eggs, beaten	3 个鸡蛋,打散
250 g dry bread crumbs	250 g 干面糠

cook knife	西餐刀
mixing bowl	拌菜碗
deep-frier	深油炸锅
skimmer	漏勺

制作方法 Method

(1) Cut chicken into 6 pieces and marinate them with salt and white pepper powder.

将鸡肉切成六份后腌制。

(2) Prepare breading:seasoned flour, egg wash, and crumbs.

准备面包糠炸。

(3) Pass the chicken through the standard breading procedure.

将鸡肉拍面粉、蘸鸡蛋、挂面包糠。

(4) Heat the fat in a deep fryer to 165-175 ℃. Fry the chicken until golden brown and cooked through.

将鸡肉放入165～175 ℃的油温中炸至成熟、金黄色即可。

(5) Remove the chicken from the fat, drain well, and serve immediately.

沥油,端盘服务。

 特点 Characters

Soft inside and crisp outside　　　　　　外酥内软

图8-1

二、煎(pan frying)

(一)概念

煎是指将加工成型的排、块、片等软嫩原料经腌制入味,后使用少量的油脂加热至上色,并达到规定火候的烹调方法。

Pan frying is a form of frying characterized by the use of minimal cooking oil or fat. In the case of a greasy food such as bacon, no oil or fat may be needed. Pan frying can retain the moisture in foods, it is used for larger pieces of food, such as chops and chicken pieces.

(二) 特点

(1) 煎制的烹调方式的用油量为原料厚度的 1/5～1/2。

(2) 煎制的烹调方式使菜品能在短时间里达到外焦里嫩的效果,与炸的烹调方式相比,煎制食物的吸油量更少,且吃起来不会太油腻。

(3) 因为煎制的烹调方式要求原料在短时间里成熟,并保持质地鲜嫩,所以并不是所有原料都适合这种烹调方法。优质、鲜嫩、水分含量高、块状或片状的肉类比较适合煎制。

(三) 种类

(1) 清煎:将鲜嫩的动物性原料改刀后,经调味、腌制,再直接放入油脂中煎制到所需火候。例如:煎牛排。

(2) 干煎:将鲜嫩的原料改刀后,经调味、腌制,蘸面粉入锅煎,用油煎至两面呈淡黄色。例如:煎鹅肝、煎三文鱼等。

(3) 软煎:又称蛋煎,将原料改刀、调味后,先蘸面粉,再蘸鸡蛋液放入煎锅中,煎至两面呈金黄色。例如:软煎鱼排。

(4) 吉列煎:又称面包糠煎,将原料改刀,经调味后先挂面粉,蘸鸡蛋液,再拍面包糠入锅中煎制。例如:吉列鸡排、吉列鱼排等。

(四) 操作要点

(1) 选用优质、鲜嫩的原料,将原料切成薄大片,提前腌制入味。

(2) 提前将煎锅刷洗干净,待锅烧热后倒入油脂涂匀。油量不宜过多,最多只能浸没原料的 1/2。

(3) 根据菜品的要求选择不同的火候。例如:煎鸡蛋用低油温下锅,煎牛排则用高油温下锅。

(4) 煎的温度一般为 120～175 ℃,不应超过 190 ℃,不低于 90 ℃。

(5) 煎制时,不能用具按压原料,避免水分流失。

(6) 若煎制的原料厚、大,短时间内不易成熟,煎后可以放入烤箱,使其成熟。

(五) 煎和炒的区别

(1) 煎制的食物一般比炒制的食物更大、更厚,烹调的温度更低,烹调的时间更长。

Pan-fried items are generally larger and thicker than sauted items, they must be cooked over lower heat for longer time, so that the surface will not become too browned before the items are cooked.

（2）煎蔬菜时，一般使用面包糠煎，避免水分的流失。

Pan-fried vegetables are often given a coating, such as breading, that gives the cooked vegetable a crisp exterior contrast with the tender vegetable inside.

（3）炒制食物需要的烹调时间短，用油少，油温高。

Saute means cooking quickly in a small amount of fat. The product is often tossed or flipped in the pan over high heat.

（4）煎制食物需要的烹调时间长，用油多，油温较低。

Pan-frying means cooking in a larger amount of fat for a longer time at lower heat, and the product is not tossed or flipped.

（六）制作实例

煎茄子配番茄少司(pan-fried eggplant with tomato sauce)

 原料 Ingredients

0.5 kg eggplant	0.5 kg 茄子
30 g flour	30 g 面粉
1/4 teaspoon of salt	1/4 茶匙盐
white peppercorn as needed	适量白胡椒
100 mL egg wash	100 mL 鸡蛋液
100 g bread crumbs	100 g 面包糠
oil for frying as needed	适量用于煎炸的油
250 mL tomato sauce	250 mL 番茄少司

 工具 Tools

iron skillet or saute pan	铁制煎锅或炒锅
slotted spatula	漏铲

 制作方法 Method

（1）Wash and trim the eggplants. Cut crosswise into 0.5 cm slices.

清洗并修整茄子，斜刀切成 0.5 cm 厚的片。

（2）Keep the eggplants in salted cold water for about 30 minutes. (It helps to prevent darkening and eliminating some bitter flavors.)

将茄子浸泡在盐冷水中约 30 min。(可以避免茄子变黑并能消除一些苦味。)

(3) Set up breading steps, season the flour with salt and white peppercorn.

准备好裹面包糠的步骤,在面粉里加入盐和白胡椒调味。

(4) Drain the eggplants and dry them well. Pass through standard breading procedure.

捞出茄子沥干,再将它们按标准方法裹上面包糠。(按面粉、鸡蛋液、面包糠的顺序给茄子逐步裹上面包糠,并用手将面包糠按压紧实。)

(5) Heat 0.5 cm oil in a heavy iron skillet or a saute pan. Pan-fry the breaded eggplants on both sides until golden brown. Remove them from pan with slotted spatula and drain them on absorbent paper.

在厚铁制煎锅或炒锅内预热约 0.5 cm 的油。放入裹好面包糠的茄子并煎炸它的两面直至呈金黄色。用漏铲将茄子从锅中捞出并放在吸油纸上吸去多余的油。

(6) Serve 2-3 slices per portion, depending on their size. And pour 60 mL tomato sauce on each portion. Ladle the sauce in a straight line across the eggplants, do not cover them completely.

上菜时根据茄子片大小,每份放 2~3 片的茄子。每份浇上 60 mL 的番茄少司。用勺将酱汁垂直于茄子浇上,不要完全覆盖住茄子。

 特点 Characters

Soft inside and crisp outside, golden colour　外酥内软,色泽金黄

三、炒(saute)

(一) 概念

炒是指把加工成型的原料条、丝、片、丁、粒放入炒锅中,再加入少量的油,用较高的温度,在短时间内不断翻炒,将原料烹调成熟的方法。

Saute means cooking quickly in a small amount of fat. High heat is required, and the procedure is most often done in a broad, flat pan called a saute pan.

(二) 特点

(1) 原料的形状较小,用旺火加热能使其快速成熟,适合质地鲜嫩的原料和一些熟料,例如:米饭、面条、蔬菜等。

(2) 炒制过程中一般不用添加过多汤汁,菜品具有鲜香脆嫩的特点。

（三）制作步骤

（1）Collect all equipments and food supplies. 准备原料和器具。

（2）Foods should be seasoned before cooking, this should be done at the last minute before cooking. 食物在炒制前应调味。

（3）Place the saute pan over high heat until the pan is hot. 大火预热炒锅。

（4）Add just enough fat to the hot pan to cover the bottom with a thin coating. 加入少量油到油锅中。

（5）Add the meat to the pan. Do not overfill the pan. 将肉放入炒锅中，不要将炒锅塞得过满。

（6）Saute meat until the first side is browned, then turn the meat over to brown the second side. 炒肉时，当一面上色后翻炒另一面。

（7）Remove the items from the pan and keep them warm. 将食物从锅中盛出来，保温。

（8）Drain excess fat. 去除多余的油脂。

（四）操作要点

（1）Use only tender materials for sauteing.
使用质地鲜嫩的原料。

（2）Smaller or thinner pieces of meat require higher heat.
形状小、薄的肉类应用较高的油温。

（3）If large or thick items are browned over high heat, it may be necessary to finish them at lower heat or in an oven to avoid burning them.
形状厚、不易成熟的原料经高油温炒制上色后，有必要调低火力或将原料放入烤箱中避免其颜色过深。

（4）Do not overfill the pan, this will lower the oil temperature.
不要将炒锅装得太满，这样会降低油温。

（5）Use clarified butter or oil or a mixture of the two for sauteing. Whole butter burns easily.
使用清澈的油脂，不宜使用黄油来炒原料。

（6）When sauteing the food, the temperature of the oil is 150-195 ℃, the amount of the oil is 1/4-1/3 of the food.
油温控制在150～195 ℃，油量为原料的1/4～1/3。

（五）制作实例

1. 炒笋瓜（saute zucchini）

原料 Ingredients

0.6 kg zucchini	600 g 笋瓜
50 mL olive oil	50 mL 橄榄油
60 g shallots or onions, minced	60 g 洋葱或青葱,切碎
1-2 garlic cloves, chopped	1~2 瓣大蒜,切碎
chopped parsley	切碎的欧芹
salt	盐
white pepper powder	白胡椒粉

 工具 Tools

saute pan	炒锅
wooden spatula	木铲

 制作方法 Method

(1) Wash and trim the zucchini. Cut crosswise into thin slices.

清洗并修整笋瓜,斜刀切成薄片。

(2) Heat the oil in the saute pan, add the shallot or onion and the garlic, saute until soft but not browned.

在炒锅中加入油,后加入青葱或洋葱、大蒜,炒至变软但不要变成棕色。

(3) Add the zucchini and saute until slightly browned but still crisp.

加入笋瓜继续炒直到轻微上色,外皮松脆。

(4) Add the parsley and toss to mix. Season to taste.

加欧芹到锅中,后加入盐和胡椒调味。

2. 炒抱子甘蓝核桃(brussels sprouts with walnuts)

原料 Ingredients

0.5 kg brussels sprouts	500 g 抱子甘蓝
20 g butter	20 g 黄油
60 g walnut pieces	60 g 核桃碎
salt	盐

工具 Tools

saute pan 炒锅
wooden spatula 木铲

 制作方法 Method

(1) Trim the base of the brussels sprouts and remove any damaged leaves.

整理抱子甘蓝的底部,去除杂叶。

(2) Blanch the brussels sprouts in a large quantity of boiling salted water.

在沸腾的盐开水中煮抱子甘蓝。

(3) Drain and refresh the brussels sprouts in ice water. Drain again.

将抱子甘蓝沥干后过冰水冲,后沥干。

(4) Cut the brussels sprouts in half lengthwise.

纵向对切抱子甘蓝。

(5) Heat the butter in a saute pan which is large enough to hold the brussels sprouts in a thin layer.

炒锅中融化黄油。

(6) Add the brussels sprouts and the walnuts to the pan. Saute until the brussels sprouts are tender and lightly browned.

加入抱子甘蓝、核桃到锅中,炒至抱子甘蓝软化、轻微变色。

(7) Add salt to taste.

加盐调味。

第二节 用水传热的烹调方法

用水进行传热的烹调方法主要是利用水或水蒸气作为热载体或传热介质,而食物原料不直接接触热源,具体的烹调方法有温煮、沸煮、蒸、烩、焖等。

Moist-heat methods are those in which the heat is transferred to the food product by water or water-based liquids such as stock and sauce, or by steam.

一、温煮(poaching)

(一)概念

温煮是指把食物原料浸入水或基础汤等液体中,用低于沸点的温度(一般为 75~90 ℃),把原料煮制成熟的烹调方法。

Poaching is a type of moist-heat cooking method that involves cooking by submerging food in a liquid, such as water, milk, stock or wine.

It uses low temperature which is suitable for delicate food, such as eggs, poultry, fish and fruit. Poaching is often considered as a healthy method of cooking because it does not use fat to cook or flavor the food.

(二)特点

温煮采用的温度比较低,因此对原料的组织及营养素的破坏较小,能更好地保持菜肴的水分,使菜肴质地鲜嫩,口味清淡,保证了食材的原汁原味。

The poaching method is used to gently cook tender poultry in order to retain moisture and develop a light, subtle flavor.

温煮时一般使用高汤,可加入酒或其他调味料。

The cooking liquid is usually stock, sometimes with the addition of wine and other flavorings and seasonings.

(三)适用范围

温煮适宜制作质地鲜嫩、粗纤维较少、水分充足的原料,如:鸡蛋、鱼虾、水果等。

(四)操作要点

(1)水或基础汤的用量应适当,以刚浸没原料为宜。

(2)烹调过程中应始终保持火候均匀一致,以使原料在相同的时间内同时成熟。

(3)烹调过程中可加盖保温,但要适当打开锅盖,让原料中的不良气味挥发出去。例如:煮鱼时,适时揭盖、加盖可以去腥、增鲜。

(4)根据原料的不同,将水或基础汤加热到 100 ℃,然后降低至 70~90 ℃,一般情况下,若原料的质地越嫩,体积越小,使用的温度越低。

(五)质量标准

(1) Foods should be cooked to the correct degree of doneness.

食物应烹制到正确的成熟度。

Poached tender poultry items and whole fish should be fully cooked but still moist and tender, not be overcooked or dry.

温煮家禽或鱼时应使其全熟,但不应过度烹制。

(2) Flavoring ingredients in the poaching liquid should enhance the flavor of the main item, not cover it up or conflict with it.

温煮液体里添加的原料应能加强主要食材的香味,而不能盖住其香味或与其香味产生冲突。

(3) After cooking, the item should be well drained. Poaching liquid should not be transferred to the plate.

烹制完后,应过滤食材,温煮的液体不能装入盘中。

(4) Sauces should enhance the flavor of the item, not cover it up. Most poached items have delicate flavor, so they should not be served with strongly flavored sauces.

少司汁应加强食物的香味,而不是掩盖食物的香味。由于大部分温煮食物的香味比较弱,因此不能搭配重口味的少司汁。

二、沸煮(boiling)

(一)概念

沸煮是指将食物原料放入液体中加热至沸腾,再用微沸的液体将原料煮至成熟的方法。

Boiling is the method of cooking food in boiling water or other water-based liquids such as stock or milk.

(二)沸煮的传热介质

(1) 含盐热水:适用于煮脂肪多、异味重的原料,例如:动物内脏。

(2) 基础汤:多指肉类基础汤(牛肉基础汤,鸡肉基础汤等),适用于煮肉类、家禽类。

(3) 菜汤:多采用新鲜蔬菜如番茄、洋葱、西芹等,调味料如柠檬汁、醋、葡萄酒等,加清水制成,适用于煮鱼类等原料。

(4) 牛奶:适用于煮制面食及谷类原料等。

(三)特点

沸煮的适用范围广泛,一般的蔬菜、禽类、肉类、豆类等原料都可以采用此种烹调方法。

沸煮的菜品具有清淡爽口的特点,同时充分保留了原料的本味,对原料营养成分的破坏较少。

Foods are suitable for boiling include vegetables, starchy foods such as rice, noodles and potatoes, eggs, meat, sauces, stocks, soups. As a cooking method, it is simple and suitable for large-scale cookery.

(四) 操作要点

(1) 原料要完全浸没于汤液中。
(2) 要及时去除汤中的浮沫。
(3) 煮制过程中一般不需加盖。
(4) 沸煮汤液的温度要达到沸点。

三、烩(stewing)

(一) 概念

烩是指把加工成型的原料,放入用本身原汁调成的浓少司中,加热至成熟的烹调方法。

Stewing is the method of cooking solid food ingredients that have been cooked in liquid and served in the resultant gravy.

Ingredients in the stewing include any combination of vegetables or meat, while water can be used as the stew-cooking liquid, wine, stock and beer are also common. Seasoning and flavourings may also be added.

(二) 烩的类型

根据在烩的过程中使用的少司不同,烩又可以分为红烩、白烩、黄烩等。
(1) 红烩:以布朗少司或番茄少司为基础,例如:红烩牛肉。
(2) 白烩:以白少司或奶油少司为基础,例如:白汁烩鸡。
(3) 黄烩:以白少司为基础,加入奶油、蛋黄等,例如:黄汁烩鸡。

(三) 特点

烩制的菜肴因为使用了原汁调成的浓少司,所以具有原汁原味、汁多味浓、色泽美观的特点。适合烩制的原料范围广泛,包括各种动物性原料、植物性原料。

Ingredients in the stewing include any combination of vegetables, such as carrots, potatoes, beans, peppers and tomatoes or meat, especially tougher meat are suitable for slow-cooking, such as beef and poultry. While water can be used as the stew-cooking liquid, wine, stock, and beer are also common. Stews are typically cooked at low temperature.

(四) 操作要点

(1) 少司以刚好覆盖原料为宜。

(2) 在烩制过程中需要加盖,避免水分蒸发过多。

(五) 制作实例

现介绍烩牛肉(beef stew)的制作。

 原料 Ingredients

500 g cubed beef	500 g 切块的牛肉
1.5 tablespoon of vegetable oil	1.5 汤匙蔬菜油
2 cubes of beef bouillon, crumbled	两块牛肉精,碎的
500 mL water	500 mL 水
1/2 teaspoon of dried rosemary	1/2 茶匙干制迷迭香
1/2 teaspoon of dried parsley	1/2 茶匙干制欧芹
1/4 teaspoon of black pepper powder	1/4 茶匙黑胡椒粉
1 large potato, peeled and cubed	1 个大土豆,去皮,切块
2 carrots, cut into 1 inch pieces	2 根胡萝卜,切成约 2.5 cm 大小的块
2 stalks celery, cut into 1 inch pieces	2 根西芹,切成约 2.5 cm 大小的块
1 small onion, chopped	1 个洋葱,切碎
1 teaspoon of cornstarch	1 茶匙淀粉
1 teaspoon of cold water	1 茶匙冷水

 工具 Tools

pot	罐子
wooden spatula	木铲

 制作方法 Method

(1) In a large pot, cook beef in oil over medium heat until brown. Dissolve bouillon in water and pour into the pot. Stir in rosemary, parsley and black pepper powder. Bring to a boil, then reduce heat, cover and simmer for 1 hour.

在一个大罐子里加入油,用中火将牛肉煎上色。待牛肉精在水中融化后倒入罐中,搅拌加入干制的迷迭香、欧芹、黑胡椒粉,煮沸后,将火调小,盖上盖子炖 1 h。

(2) Stir potatoes, carrots, celery, and onion into the pot. Dissolve cornstarch in 1 teaspoon of cold water and stir into stew. Cover and simmer for 1 hour.

搅拌加入至罐中的土豆、胡萝卜、西芹和洋葱。将淀粉溶解后倒入混合物中,盖上盖子,再炖 1 h。

四、焖(braising)

(一)概念

焖是指把加工成型的原料,经初步热加工后,放入少量的汤水中使之成熟的烹调方法。焖制一般在密闭的容器内进行。

Braising is a combined cooking method that uses both wet and dry heats, typically, the food is first seared at high temperature, then finished in a covered pot at lower temperature while sitting in the liquid.

The basic principle of braising is the combination of searing or browning and then simmering.

(二)特点

焖制菜肴需要的加热时间较长,菜品具有软烂、味浓、原汁原味的特点。

主要适用于结缔组织较多的原料,焖制时间根据原料的质地而定。

焖肉类和禽类菜肴时,首先要使原料上色,然后将原料放在汤汁中炖熟。

To braise the meat and poultry, first brown or sear them in fat or in a hot oven, then simmer in a flavorful liquid until tender.

焖制的肉类的质量很大程度上取决于汤汁的质量。其他常用的焖汁包括葡萄酒、腌汁等。

The quality of braised meat depends greatly on the quality of the stock that the meat is cooked in. Other liquids used in braising include wine, marinades and so on.

(三)操作要点

(1) 汤汁的用量应适当,一般以浸过原料的 1/3~1/2 为宜。

(2) 焖制前,要对原料进行初步处理。例如:大块肉类可提前腌制入味,使肉质松软;也可以用热油将大块肉类煎上色,使其表面结成硬壳,保持水分。

(3) 焖制时,容器要加盖。

(四)质量标准

(1) Braised items should be fork-tender. The interior should be juicy.

焖制的菜肴软烂,内部多汁。

(2) Although tender, braised foods should maintain their shapes and not be falling apart.

焖制的菜肴需保持完整的外观。

(3) Braised foods should have well-developed, concentrated flavor due to long cooking in flavorful liquids.

焖制的菜肴风味浓郁。

(4) Braised foods should have an attractive color developed from proper browning or searing at the beginning of the cooking process.

焖制的菜肴有着诱人的色泽。

五、蒸(steaming)

(一) 概念

蒸是指将原料加工成型、调味后,放入容器内,利用蒸汽加热,使原料成熟的烹调方法。

Steaming means to cook foods by exposing them directly to steam. In a large quantity of cooking, this is usually done in special steam cookers.

Steaming is a popular method for cooking vegetables. In Western cuisines, it is less often used for meats, poultry, and fish, although it is more common in some Asian cuisines. This cooking method is best suited for foods that are naturally tender.

(二) 特点

蒸制菜肴时用油较少,又是在密闭的容器内加热,因此蒸制的菜肴口感清淡,原汁原味,能更好地保持原料的造型,在烹制过程中的营养素损失很少。

适用于制作质地鲜嫩、水分充足的原料,例如:鱼虾等。

(三) 操作要点

(1) 蒸制前要先对原料进行调味。
(2) 蒸制容器在加热过程中要密封好。
(3) 蒸制时要根据不同的原料来掌握火候。

(4) 蒸制菜肴时应防止蒸汽烫伤。

(四) 质量标准

(1) Items should be cooked just to expected doneness and be moist and tender.

原料应烹制到预期的成熟度,鲜嫩多汁。

(2) Items appear plump, not shrunken, and the surface is free of coagulated albumen.

菜品的外观饱满,表面无蛋白凝固现象。

(3) Items should have good natural color.

菜品颜色自然。

(4) Flavors should be delicate.

香味自然。

第三节 用空气传热的烹调方法

用干热空气传热的烹调方法在西餐烹调中的使用是非常广泛的。传热介质是热空气、油脂和金属铁板。传热的形式是对流、辐射和传导。用热空气传热的烹调方法使用的温度范围广,低温可在100 ℃以下,最高温度可以达到300 ℃以上。烹调方法包括烤、焗、铁扒、串烧等。

一、烤(roasting)

(一) 概念

烤是指将经过加工成型的大块原料(如整条、整只)或小型原料(如排、片、块)等,经调味和长时间腌制后,入烤炉前在原料表面浇上少量的油脂,放入封闭的烤炉中,利用高温热空气的辐射、对流,使原料上色并达到所需火候要求的烹调方法。

Roasting is a cooking method that uses dry heat where hot air envelops the food, and all sides of the food are cooked evenly with temperature of at least 150 ℃. Roasting can enhance flavor through caramelization and Maillard browning on the surface of the food.

（二）特点

烤适用于加工制作动植物原料的鲜嫩部位。烤制的菜品外部油润光亮，外焦里嫩。

（三）适用范围

烤的烹调方法适用范围较广，适用于加工制作各种形状较大的肉类原料（T骨牛排、羊腿等）、禽类原料、野味及蔬菜和部分面点制品。

Roasting uses indirect, diffused heat, and is suitable for slower cooking of meat in a larger, whole piece. Meats, most root and bulb vegetables can be roasted. Any piece of meat, especially red meat, that has been cooked in this fashion is called a roast. Meats and vegetables prepared in this way are described as "roasted".

（四）基本流程及操作要点

（1）初加工原料，去除多余的脂肪和结缔组织。有些原料需要用细绳捆扎，防止在烤制过程中变形。

（2）腌制原料，使其渗透入味。

（3）将烤炉预热至220 ℃，待原料入炉后能使表面迅速上色、凝结，防止水分的过多流失，再把温度降至150～180 ℃，烤至所需要的火候。

（4）烤制质地鲜嫩、水分充足、易成熟的原料时，应直接采用220～240 ℃的高温使其快速达到所需的火候，防止水分的过多流失。

（5）蔬菜、香料不要切得太小，否则在高温下容易烤焦。

（6）烤制过程中，应不断往原料上浇烤肉汁，以增加菜品的滋味。

（7）烤制过程中如发现温度过高，导致菜品颜色过深，应立即降温，并在原料上覆盖一层锡纸，防止颜色继续加深。

（8）应在达到烤制的所需火候前提早出炉，因为菜品出炉后其自身所带的温度会使火候继续加大。

（9）肉类原料出炉后不能立即切割，应待肉质的内、外温度一致，汁液稳定后再切割，防止切割时汁液流出。

（五）烤制的肉类菜品成熟度的检验

（1）通过感官检验，观察肉类原料外观的收缩率。

（2）用肉针扎菜品检验，通过流出来的血水来判断：汁液为鲜红色说明菜品是生的；汁液为淡红色说明已基本成熟；汁液为白色说明已完全成熟。但不宜过多扎试，以免肉汁流失。

（3）使用温度计测试。以红肉为例，使用温度计扎入肉质内部显示30℃，说明烤肉达到三成熟(rare);显示 40 ℃说明烤肉已达到四成熟(medium rare);显示 50℃时说明烤肉已达到五成熟(medium);显示 70 ℃时说明肉质已全熟(well done)。

(4) 用手指按压检验。三四成熟的烤肉肉汁较多,肉质松软,无弹性;五六成熟的烤肉弹性较小;七八成熟的烤肉弹性较大,肉质较硬;全熟的烤肉肉质坚实,弹性强。

(六) 制作实例

现介绍烤牛肉配青葱、蘑菇(roast fillet of beef with shallots and mushrooms)的制作。

 原料 Ingredients

2 tablespoons of soy sauce	2 汤匙酱油
1 tablespoon of red wine vinegar	1 汤匙红酒醋
6 tablespoons of olive oil	6 汤匙橄榄油
8 garlic cloves, peeled,4 crushed,4 left whole	8 瓣大蒜,去皮,4 瓣压碎,4 瓣为整个
1 beef fillet, about 1.75 kg	1.75 kg 牛柳
10 small shallots, peeled	10 颗青葱,去皮的
a small bunch rosemary, stems removed	1 小束迷迭香,去除根部
350 g mushrooms, halved or quartered	350 g 的蘑菇,对半切或每个切成四份

 工具 Tools

frying pan	煎锅
roasting tray	烤盘

 制作方法 Method

(1) Mix 1 tablespoon of soy sauce, the vinegar, and 3 tablespoon of olive oil with 4 cloves garlic, crushed. Set aside. Season the beef with plenty of salt and mixture, then put in a large plastic zip bag with the marinade. Marinate in the fridge for at least 1 hour and up to 24 hours.

将 1 汤匙酱油、1 汤匙红酒醋、3 汤匙橄榄油和 4 瓣压碎的大蒜混合,静置。将牛肉用盐及大蒜的混合物腌制后,装入封闭的塑料袋中,后放入冰箱冷藏 1～24 h。

(2) Leave the beef at room temperature for a couple of hours before roasting. Pat dry and season again. Heat the oven to 200 ℃. Heat 1 tablespoon of olive oil in a frying pan and sear the beef well on all sides, then put it in a large roasting tray with

the shallots. Sprinkle over half the rosemary and roast for 25 minutes for rare beef(give it another 10 minutes for medium), cover it with foil.

烤制前将牛肉放置室温下 2 h,拍干、调味。将烤箱预热到 200 ℃,在锅中加入 1 汤匙橄榄油,将牛肉两面煎上色后,连同青葱放入烤盘中。撒入一半的迷迭香后盖上锡纸烤 25 min(三成熟牛肉),再烤 10 min(牛肉五成熟)。

(3) Mix the mushrooms with the remaining olive oil, soy sauce, garlic cloves and rosemary. Put the mixture on a large shallow baking tray, season and roast it in the oven for 15 minutes. Slice the meat and serve with a spoonful of mushrooms and shallots.

将蘑菇和剩下的橄榄油、酱油、大蒜、迷迭香混合后,放入浅的烤盘中,调味后烤制 15 min。将牛肉切片并搭配一勺蘑菇和青葱食用。

二、焙烤(baking)

(一)概念

将原料放入烤炉内,以热空气对流和辐射换热的方式加热食物的方法统称为焙烤。主要用于烤制面包、糕饼、饼干等。

Baking is a method of cooking food that uses prolonged dry heat, normally in an oven, but also in hot ashes, or on hot stones. The most common baked item is bread, but many other types of foods are baked.

(二)特点

焙烤制品要严格按照配方来计量,成品呈金黄色,色泽诱人。

(三)基本流程及操作要点

(1) 大多数的焙烤制品需要用专门的模具盛装。将模具先刷油,后装入生胚,最后放入烤炉烤制。

(2) 烤炉应提前预热,严格按照要求的温度和时间烤制,具体的温控和时间要看品种的规定和要求。

(3) 蔬菜、肉类原料在烤制时应先用锡纸和面皮包裹,然后再入炉烤。

(4) 部分产品应用隔水加热的方法烤制,以利于蛋白质的凝固,例如:布丁。

(5) 烤和焙烤的区别:烤需要往原料表面浇烤肉汁,焙烤则不需要。

三、烧烤(barbecuing)

（一）概念

烧烤，又称明火烤，是将鲜嫩的动物性和植物性原料，经过加工、调味腌制后用串针串起，刷上油脂，放在专用的木炭明火炉上，不断转动串针，通过炙烤使原料加热成熟的方法。

Barbecuing means to cook with dry heat created by the burning of hardwood or coals.

（二）特点

味道浓郁，焦香。

（三）适用范围

适用于制作质地鲜嫩的原料，如：鸡肉、里脊肉、鸡肝等，新鲜、鲜嫩的蔬菜。

（四）基本流程及操作要点

(1) 选用鲜嫩的原料，并提前腌制。
(2) 烧烤时要待木炭烧尽、无烟时进行。
(3) 烤制大块的肉排、鱼片时不用串针，使用烤架即可。
(4) 不要将原料穿得过紧，应尽量平整，以便于加热。

四、焗烤(gratin)

（一）概念

焗烤又称面火烤，是指将加热成熟的菜品，上面浇上一层高油脂、浓稠的少司或奶酪后，放入明火焗炉中，利用高温热空气把菜品表面焗成金黄色的烹调方法。

Gratin is a widespread used culinary technique in which an ingredient is topped with a browned crust, often using bread crumbs, grated cheese, egg and butter. Gratin originated from French cuisine, it is baked or cooked under an overhead grill or broiler to form a golden crust on top and is traditionally served in its baking dish.

（二）特点

高温加热菜品后，菜品的表面形成一层硬皮，可以有效地保护菜肴，避免水分的流失,保证原料的鲜嫩,并增加菜肴的美观,同时具有气味芳香、色泽金黄、香味浓郁的特点。

（三）适用范围

主要适用于加工制作肉质鲜嫩的鱼类、肉类及蔬菜等,多用于意式面食的制作。

（四）基本流程及操作要点

（1）焗盘内需要涂上黄油,预防菜品粘连。要保证菜品表层的少司等厚薄均匀、平整。

（2）焗烤的温度比较高,一般为180～300 ℃。

（3）烤炉内加热以面火为主,底火为辅。

（4）菜品应迅速上色,不宜长时间加热,否则会影响菜品的质量。

（5）菜品出炉时应注意安全,以防烫伤。

五、铁扒(grilling)

（一）概念

铁扒是西餐烹调方法中比较有特色的一种,原料经加工成型、调味抹油后,放在扒炉上,利用炉条的传导和辐射,在原料表面烙上焦黄色、均匀的烙纹,并达到所需火候要求的烹调方法。

Grilling is a method of cooking that involves dry heat applied to the surface of food, commonly from above or below. Grilling usually involves a significant amount of direct, radiant heat, and tends to be used for cooking meat quickly.

Food to be grilled is cooked on a grill, a grill pan or griddle.

（二）特点

烹调温度高,使原料表层迅速炭化,原料内部的水分流失较少,口感焦香,鲜嫩多汁,纹路规则。

（三）适用范围

由于铁扒是一种温度高、时间短的烹调方法,适用于制作质地鲜嫩、质量上乘的肉类原料,例如:T骨牛排、西冷牛排、鳟鱼等。

（四）基本流程及操作要点

（1）铁扒炉需提前预热,清理铁扒炉表面的烟尘并刷油。

（2）先用高温使原料上色，封闭原料的表面，防止汁水的流失，再根据需要降温。

（3）使用专用夹子翻动原料，以免破坏原料的表面。

（五）牛扒的成熟度

（1）二三成熟（rare）牛扒的表面略有焦黄色，中间是鲜红的生肉，无弹性，内部温度为 25~30 ℃。

（2）三四成熟（medium rare）牛扒的表面为焦黄色，略有弹性，中心一层为鲜红的生肉，内部温度为 35~40 ℃。

（3）五成熟（medium）牛扒的表面焦黄有弹性，中心为粉红色，内部温度为 45~50 ℃。

（4）七八成熟（medium well）牛扒的表面为焦黄色，弹性更大，中心肉色为浅红色，内部温度为 55 ℃。

（5）全熟（well done）牛扒的表面为咖啡色，硬实，内部温度为 70 ℃。

第九章 西餐素食菜品的制作工艺

第一节 素食主义概述

一、素食主义定义

素食主义是一种饮食的文化,实践这种饮食文化的人称为素食主义者(vegetarian),他们不食用有主观意识的动物,包括家畜、野兽、飞禽、鱼类等,也不食用由动物身上各部分所制成的食物,包括动物油、动物胶,但一般可以食用蛋、奶、黄油、奶酪等。

除素食主义外,还有一种纯素食主义(vegan)。和素食主义者不同的是,纯素食主义者连蛋、奶制品、蜂蜜也不能食用,即凡来源于动物的食品一律不沾。

二、素食的原因

(一)宗教因素

世界上许多素食者都是基于宗教的因素而食素。

(二)非宗教因素

因为传染性疾病等环境因素、道德因素,或因保护动物、保证健康、减肥等而吃素。

第二节 西餐素食菜品制作

一、小食类(snack)

(一)蒜蓉面包片(garlic bread)

 原料 Ingredients

120 g butter, melted	120 g 融化的黄油
1 teaspoon of garlic salt	1 茶匙大蒜盐
1/4 teaspoon of dried rosemary	1/4 茶匙干制迷迭香
1/4 teaspoon of dried basil	1/4 茶匙干制罗勒
1/4 teaspoon of dried thyme	1/4 茶匙干制百里香
1/8 teaspoon of garlic powder	1/8 茶匙大蒜粉
1 tablespoon of grated Parmesan cheese	1 汤匙帕玛森奶酪,磨碎
2 baguettes, sliced	2 根法棍面包,切片

 工具 Tools

bowl	碗
baking pan	烤盘
small knife	小刀
oven	烤箱

 制作方法 Method

(1) Preheat oven to 150 ℃.
预热烤箱到 150 ℃。

(2) In a small bowl, mix butter, garlic salt, rosemary, basil, thyme, garlic powder and Parmesan cheese.
在小碗里混合黄油、大蒜盐、迷迭香、罗勒、百里香、大蒜粉和帕玛森奶酪。

(3) Spread each slice of baguette with equal portion of the butter mixture. Sprinkle with additional Parmesan cheese, if desired.

在切好的法棍面包上均匀地涂抹黄油混合物，撒上帕玛森奶酪。

(4) Place bread on a medium baking sheet. Bake in the preheated oven for 10-12 minutes, or until the edges are very lightly browned.

将面包放置在中等大的烤盘中，烤制 10～12 min 直至轻微上色。

（二）菠菜蛋糕（spinach cake）

 原料 Ingredients

250 g spinach, rinsed and chopped	250 g 菠菜，洗净切碎
250 g all-purpose flour	250 g 普通面粉
1 teaspoon of salt	1 茶匙盐
1 teaspoon of baking powder	1 茶匙发酵粉
2 eggs	2 个鸡蛋
250 mL milk	250 mL 牛奶
120 g butter, melted	120 g 融化的黄油
1 onion, chopped	1 个洋葱，切碎
250 g shredded Mozzarella cheese	250 g 切碎的马苏里拉奶酪

 工具 Tools

saucepan	少司锅
bowl	碗
baking pan	烤盘
oven	烤箱
spatula	铲子

 制作方法 Method

(1) Preheat oven to 190 ℃. Lightly grease the baking dish.

预热烤箱到 190 ℃，在烤盘上刷油。

(2) Place spinach in a medium saucepan with enough water to cover. Bring to a boil. Simmer with lower heat and cook until spinach is limp for about 3 minutes. Remove from heat, drain, set aside.

将菠菜放入装有水的、中等大的少司锅中，煮沸后用小火炖，直至菠菜变得松软，约

3 min,离火,沥干水分,静置。

(3) In a large bowl, mix flour, salt and baking powder. Stir in eggs, milk and butter. Mix in spinach, onion and Mozzarella cheese.

在大碗中混合面粉、盐、发酵粉,后加入鸡蛋、牛奶、黄油,最后加入菠菜、洋葱和马苏里拉奶酪混合。

(4) Transfer the mixture to the prepared baking dish. Bake in the preheated oven for 30 to 35 minutes, cool before serving.

将混合物倒入预先准备的烤盘中,烤制 30～35 min。

(三) 牛油果小食(stuffed avocados with corn and olives)

100 g lightly cooked corn kernels　　　　100 克熟玉米粒
60 g finely diced yellow bell pepper　　　60 g 黄灯笼椒,切丁
1 medium tomato, finely diced　　　　　1 个中等大的番茄,切丁
60 g chopped pitted black olives　　　　 60 g 去核黑橄榄,切碎
2 tablespoons of minced fresh cilantro　　2 汤匙切碎的新鲜芫荽叶
juice of 1/2 lemon　　　　　　　　　　　半个柠檬,榨汁
salt and freshly ground black pepper　　　盐和现磨黑胡椒
2 medium firm, ripe avocados　　　　　　2 个中等大、成熟的牛油果

mixing bowl　　　　　　　　　　　　　拌菜碗
spoon　　　　　　　　　　　　　　　　勺子

(1) Combine all the ingredients except the avocados in a mixing bowl.
混合除了牛油果外的所有原料。

(2) Cut the avocados in half, remove the pit, scoop out some of the pulp, leaving sturdy shells about 1/4 inch thick and fill them with the mixture(chop the avocado flesh and stir into the stuffing mixture).

将牛油果对半切开,去除核,挖出部分果肉后,将果肉拌入混合物,最后在牛油果中间填充混合物。

二、汤类(soup)

(一)奶油芦笋汤(cream of asparagus soup)

 原料 Ingredients

800 g asparagus	800 g 芦笋
1/2 tablespoon of olive oil	半汤匙橄榄油
1 large onion, chopped	1个大洋葱,切碎
1 clove garlic, minced	1瓣大蒜,剁碎
1 large potato, finely diced	1个大土豆,切丁
1 L vegetable broth	1 L 蔬菜汤
50 g chopped fresh dill	50 g 切碎的莳萝
1 teaspoon of dried basil	1茶匙干制罗勒
pinch of nutmeg	少量肉豆蔻
200 mL milk	200 mL 牛奶
salt and peppercorn	盐、胡椒

 工具 Tools

vegetable peeler	去皮器
soup pot	汤锅
blender	搅拌器
slotted spoon	漏勺
wooden spatula	木铲

 制作方法 Method

(1) Cut about 1 inch off the bottoms of the asparagus stalks and discard. Scrape any tough skin with a vegetable peeler. Cut the stalks into approximately 1 inch pieces, set aside the tips.

将芦笋茎部底端约1寸(3.33 cm)处去除,刮去粗糙的外皮,将芦笋切成约1寸(3.33 cm)大小。

(2) Heat the oil in a large soup pot. Add the onion and garlic, and saute until golden.

汤锅中加入油后,加热,加入洋葱、大蒜炒至金黄色。

(3) Add the asparagus pieces, potato, vegetable broth, dill, basil, and nutmeg. Bring to a boil, then cover and simmer gently until the asparagus pieces and potato are tender for about 20 minutes. Remove from the heat.

后加入芦笋、土豆、蔬菜汤、莳萝、罗勒和肉豆蔻,煮沸后加盖,炖至芦笋和土豆变软,约 20 min,离火。

(4) Transfer the solid ingredients to the container of a blender with a slotted spoon, process in batches until smoothly pureed, then stir back into the liquid in the soup pot.

用漏勺将固体的原料倒入搅拌器中,搅拌直至成光滑的泥状,最后倒入汤锅中。

(5) Add enough milk to give the soup a slightly thick consistency. Season with salt and peppercorn and return to low heat.

加入足够的牛奶到锅中使混合物变得浓稠,最后用盐、胡椒调味。

(二) 土豆甜玉米杂烩汤(potato and sweet corn chowder)

 原料 Ingredients

6 sweet corn cobs	6 根甜玉米棒
2 tablespoons of vegetable oil	2 汤匙植物油
1 onion, finely diced	1 个洋葱,切丁
2 garlic cloves, crushed	2 瓣大蒜,压碎
1 celery stalk, diced	1 根芹菜梗,切丁
1 carrot, peeled and diced	1 根胡萝卜,去皮、切丁
2 large potatoes, peeled and diced	2 个大土豆,去皮、切丁
1 L vegetable stock	1 L 蔬菜汤
2 tablespoons of finely chopped flat-leaf parsley	2 汤匙切碎的扁叶欧芹

 工具 Tools

saucepan　　　　　　　　　　　　　　少司锅
blender　　　　　　　　　　　　　　　搅拌器

 制作方法 Method

(1) Bring a large saucepan of salted water to the boil. Cook the sweet corn cobs for 5 minutes. Reserve 250 mL of the cooking water. Cut the corn kernels from the cob,

place half in a blender with the reserved cooking water, and blend until smooth.

在一个大少司锅里煮沸一锅盐水,将甜玉米棒煮 5 min,预留 250 mL 的烹调水。用刀将玉米粒从玉米棒上取下,将一半的玉米粒和预留的烹调水一起倒入搅拌机中搅打至顺滑。

(2) Heat the oil in a large saucepan. Add the onion, garlic, celery and a large pinch of salt and cook for 5 minutes. Add the carrot and potatoes, cook for another 5 minutes, then add the stock, corn kernels and blended corn mixture. Reduce the heat and simmer for 20 minutes, or until the vegetables are tender. Season well and stir in the chopped parsley before serving.

将油倒入少司锅内加热,后加入洋葱、大蒜、芹菜和一大撮盐一起炒制 5 min。然后加入胡萝卜和土豆,再煮 5 min,后倒入蔬菜汤、玉米粒和搅打好的玉米蓉,将火调小炖 20 min 或直至蔬菜变软,调味。上菜前加入切碎的欧芹并搅拌均匀。

(三)南瓜汤(pumpkin soup)

原料 Ingredients

1 tablespoon of olive oil	1 汤匙橄榄油
1 onion, finely chopped	1 个洋葱,切碎
400 g pumpkin, peeled (deseeded and chopped into chunks)	400 g 南瓜,去皮,去籽,切成块
300 mL vegetable stock	300 mL 蔬菜汤
70 mL double cream	70 mL 浓奶油
for the croutons	烤碎面包块
1 tablespoon of olive oil	1 汤匙橄榄油
2 slices of bread, crusts removed	2 片面包,去除外皮

工具 Tools

saucepan	少司锅
hand blender	手持搅拌器
frying pan	煎锅

制作方法 Method

(1) Heat the olive oil in a large saucepan, then gently cook the onions for 5 minutes, until soft but not coloured.

在少司锅内加入橄榄油,炒香洋葱直至洋葱变软,但没有上色。

(2) Add the pumpkin to the pan, then cook for 8-10 minutes until it starts to soften and turn golden.

加入南瓜到锅中,炒 8~10 min 直至南瓜变软,颜色变成金黄色。

(3) Pour the stock into the pan and season with salt and peppercorn. Bring to the boil, then simmer for 10 minutes until the pumpkin is very soft.

将蔬菜汤倒入锅中,调味后煮沸,然后炖制 10 min 直至南瓜变得很软。

(4) Pour the double cream into the pan, bring back to the boil, then puree with a hand blender.

将浓奶油倒入锅中,煮沸后,用手持搅拌器将混合物打成泥状。

(5) To make the croutons: cut the bread into small squares. Heat the olive oil in a frying pan, then fry the bread until it starts to become crisp.

制作烤碎面包块:将面包切成小方块。在煎锅中加入橄榄油,煎制面包块直至变得焦脆。

(6) Reheat the soup if needed, taste for seasoning, then serve the soup with croutons.

必要时可加热汤,调味,将烤碎面包块和汤搭配。

(四)红薯梨汤(sweet potato and pear soup)

 原料 Ingredients

25 g butter	25 g 黄油
1 small white onion, finely chopped	1 个小的白洋葱,切碎
700 g sweet potato, peeled and cut into 2 cm cubes	700 克红薯,去皮,切 2 cm 大小的块
2 firm pears, peeled, cored, cut into 2 cm cubes	2 个偏硬的梨,去皮去核,切 2 cm 大小的块
700 mL vegetable stock	700 mL 蔬菜汤
250 mL pouring cream	250 mL 稀奶油
mint leaves, to garnish	薄荷叶,用作装饰

 工具 Tools

saucepan	少司锅
blender/food processor	搅拌机/食物料理机

制作方法 Method

(1) Melt the butter in a saucepan over medium heat. Add the onion and cook for 2-3 minutes, or until soft. Add the sweet potato and pear, and cook, stir, for 1-2 minutes. Add the vegetable stock, bring to the boil and cook for 20 minutes, or until the sweet potato and pear are soft.

取一个少司锅置于中火上,并将黄油融化。加入洋葱炒制2~3 min 直至洋葱变软。加入红薯和梨,翻炒1~2 min。加入蔬菜汤煮沸后,再煮20 min直至红薯和梨变软。

(2) Cool slightly, then place the mixture in a blender or food processor and blend in batches until smooth. Return to the pan, stir in the cream and gently reheat without boiling. Season and garnish with the mint.

稍微放凉后,将混合物分批倒入搅拌机或食物料理机中搅打至顺滑。之后倒回锅中拌入奶油小火加热,但不要煮开。调味后用薄荷叶做装饰。

三、沙拉类(salad)

(一)苦苣配蒜香烤面包丁沙拉(frisee and garlic crouton salad)

 原料 Ingredients

vinaigrette	调味汁
1 French shallot, finely chopped	1颗法国青葱,切碎
1 tablespoon of Dijon mustard	1汤匙法国第戎市芥末
2 tablespoons of tarragon vinegar	2汤匙龙蒿醋
150 mL extra virgin olive oil	150 mL 特级初榨橄榄油
1 tablespoon of olive oil	1汤匙橄榄油
1/2 bread stick, sliced	1/2条面包,切片
4 whole garlic cloves	4整瓣大蒜
1 baby frisee, washed	1颗嫩叶苦苣,洗净
100 g walnuts, toasted	100 g 烤过的核桃
100 g feta cheese, crumbled	100 g 菲达奶酪,磨碎

 工具 Tools

frying pan 煎锅

制作方法 Method

(1) To make the vinaigrette, whisk together the shallot, mustard and vinegar in a bowl. Slowly add the oil, whisk constantly until thickened. Set aside.

首先制作沙拉调味汁,在一个碗中将青葱末、芥末和醋混合均匀。缓慢加入橄榄油,不断搅拌直至酱汁变稠,放置一边备用。

(2) Heat the oil in a large frying pan over medium-high heat. Add the bread and garlic cloves and cook for 5-8 minutes, or until the croutons are crisp. Remove the garlic from the pan. Once the croutons are cool, break them into small pieces.

将油倒入一个大煎锅中,中火预热。放入面包和蒜瓣煎5~8 min直至面包变得香脆。从锅中取出蒜瓣。待面包冷却后将其掰成小块。

(3) Place the frisee, croutons, walnuts, feta cheese and vinaigrette in a large bowl. Toss together well and serve.

将苦苣、烤面包丁、核桃、菲达奶酪和调味汁倒入一个大碗中,搅拌均匀后立刻享用。

(二) 鳄梨西柚沙拉(avocado and grapefruit salad)

 原料 Ingredients

2 grapefruits	2个西柚
1 ripe avocado	1个成熟的牛油果
180 g watercress leaves	180 g 西洋菜叶
1 French shallot, finely sliced	1根法国青葱,切薄片
1 tablespoon of sherry vinegar	1汤匙雪莉酒醋
3 tablespoons of olive oil	3汤匙橄榄油

 工具 Tools

salad bowl   沙拉碗

 制作方法 Method

(1) Peel and segment the grapefruit, working over a bowl to save any juice drips for the dressing.

将西柚去皮并取肉,在一个碗上进行操作以保留用于做调味汁的西柚汁。

(2) Cut the avocado into 2 cm wedges and put in a bowl with the watercress, grapefruit and shallot.

将牛油果切成 2 cm 的楔形并与西洋菜叶、西柚果肉和法国青葱一同放入碗中。

(3) Put 1 tablespoon of the reserved grapefruit juice in a small, screw-top jar with the sherry vinegar, olive oil, salt and black peppercorn, and shake well. Pour the dressing over the salad.

在一个带螺旋盖的小罐子中放入 1 汤匙预留的西柚汁、雪莉酒醋、橄榄油、盐和黑胡椒并摇匀。后将调味汁浇在沙拉上。

(三) 烤番茄意粉沙拉佐青酱(roasted tomato and pasta salad with pesto)

 原料 Ingredients

140 mL olive oil	140 mL 橄榄油
500 g cherry tomatoes	500 g 樱桃番茄
3 garlic cloves, unpeeled	3 瓣大蒜,未去皮
400 g penne	400 g 笔尖通心粉
80 g pesto	80 g 青酱
3 tablespoons of balsamic vinegar	3 汤匙意大利香脂醋
basil leaves, to garnish	罗勒叶,装饰用
salt and peppercorn	盐和胡椒

 工具 Tools

saucepan	少司锅
roasting dish	烤盘

 制作方法 Method

(1) Preheat the oven to 180 ℃. Put 2 tablespoons of olive oil in a roasting dish and place in the oven for 5 minutes. Add the cherry tomatoes and garlic to the dish, season and toss until well coated. Return to the oven and roast for 30 minutes.

烤箱预热至 180 ℃,将 2 汤匙橄榄油涂在烤盘上并放入烤箱中加热 5 min,然后放入樱桃番茄和大蒜至烤盘中,使它们均匀裹上调味料,放回烤箱中烤制 30 min。

(2) Meanwhile, cook the penne in a large saucepan of rapidly boiling water until al dente. Drain and transfer to a large serving bowl.

与此同时,将笔尖通心粉倒入装有沸水的大少司锅中煮至有嚼劲。沥干水分后倒入一个大碗中。

(3) Squeeze the flesh from the roasted garlic cloves into a bowl. Add the remaining

olive oil, the pesto, vinegar and 3 tablespoons of the tomato cooking juice. Season and toss to combine. Add to the penne and mix well, ensuring that the penne is coated in the dressing. Gently stir in the cherry tomatoes, then scatter with basil leaves. Serve warm or cold.

将烤大蒜的果肉挤入碗中,加入剩下的橄榄油、青酱、醋和3汤匙烤番茄时出的汁水,调味并搅拌均匀。加入笔尖通心粉并充分搅拌,确保笔尖通心粉完全被酱汁包裹。轻轻地拌入番茄,在沙拉上均匀铺上罗勒叶。冷热都可享用。

四、主菜类(main course)

(一)亚洲米饭(Asian rice mixture)

原料 Ingredients

300 g long-grain rice	300 g 长粒米
2 tablespoons of olive oil	2 汤匙橄榄油
1 large red onion, finely chopped	1 个大紫洋葱,切碎
2 garlic cloves, crushed	2 瓣蒜,压碎
1 tablespoon of finely chopped fresh ginger	1 汤匙新鲜生姜,切碎
1 long red chilli, seeded and thinly sliced	1 根长红辣椒,去籽切成薄片
3 spring onions, finely sliced	3 根青葱,切薄片
2 tablespoons of soy sauce	2 汤匙酱油
1/2 teaspoon of sesame oil	1/2 茶匙芝麻油
2 teaspoons of black vinegar	2 茶匙黑醋
1 tablespoon of lime juice	1 汤匙青柠汁
30 g chopped coriander leaves	30 g 切碎的香菜叶

工具 Tools

saucepan	少司锅
frying pan	煎锅

制作方法 Method

(1) Bring 1 litre of water to the boil in a large saucepan. Add the rice and cook it, uncovered, for 12-15 minutes over low heat, or until the grains are tender. Drain and

rinse the rice under cold with running water, then transfer to a large bowl.

在一个大的少司锅中将 1 L 水煮沸。加入米,不加盖,小火煮 12～15 min 直至米粒变软。沥干后用冷水冲,然后放入一个大碗中。

(2) While the rice is cooking, heat the oil in a frying pan over medium heat. Add the onion, garlic, ginger and chilli, and cook them for 5-6 minutes, or until the onion has softened, but not browned. Stir in the spring onion and cook for another minute.

与此同时,将一只煎锅置于中火上并加热油。加入洋葱、大蒜、生姜和辣椒,炒制 5～6 min 直至洋葱变软但没有上色,倒入青葱后再翻炒 1 min。

(3) Remove the onion mixture from the heat and add it to the rice with the soy sauce, sesame oil, black vinegar, lime juice and coriander, mix well. Cover the rice salad and refrigerate until you are ready to serve.

离火后将洋葱混合物和酱油、芝麻油、黑醋、青柠汁、香菜一起倒入米饭里并充分搅拌均匀。封好碗口后放入冰箱冷藏直至食用时取出。

(二)罗勒青酱比萨(basil pesto pizza)

  原料 Ingredients

60 mL basil pesto　　　　　　　　　　60 mL 罗勒青酱
2 precooked pizza crusts(8 inches)　　2 张预烤比萨饼皮(8 寸(26.67 cm))
1 sliced tomato　　　　　　　　　　　1 个番茄,切片
50 g grated Mozzarella cheese　　　　50 g 磨碎的马苏里拉奶酪
coarse salt and black peppercorn　　　粗盐和黑胡椒

  工具 Tools

baking sheet 烤盘

  制作方法 Method

(1) Preheat oven to 220 ℃. Spread basil pesto on each of two pizza crusts.
烤箱预热至 220 ℃,将两张比萨饼皮都涂上罗勒青酱。

(2) Layer each crust with tomato, sprinkle with grated mozzarella cheese. Season with coarse salt and black peppercorn.
将两张饼皮铺上番茄片,撒上磨碎的马苏里拉奶酪,用盐和黑胡椒调味。

(3) Transfer to a baking sheet, bake until cheese has melted and crust is golden brown for about 10 minutes. Slice and serve immediately.

将比萨放在烤盘上，烤至奶酪完全融化，饼皮呈金黄色，大约 10 min。取出切片后即可享用。

（三）波纹贝壳状通心粉配蔬菜酱汁（rigatoni with chunky vegetable sauce）

 原料 Ingredients

1 yellow onion, diced	1 个黄洋葱，切丁
2 zucchini, diced	2 根西葫芦，切丁
2 garlic cloves, minced	2 瓣大蒜，切碎
200 g mushrooms, trimmed and quartered	200 g 蘑菇，去根后分成四份
1 can of tomato puree	1 罐番茄酱
400 g rigatoni	400 g 波纹贝壳状通心粉
2 tablespoons of fresh oregano leaves, coarsely chopped	2 汤匙新鲜的牛至叶，粗略切碎
2 tablespoons of extra-virgin olive oil	2 汤匙特级初榨橄榄油
coarse salt and black peppercorn	粗盐和黑胡椒

 工具 Tools

saucepan　　　　　　　　　　　　　　少司锅

 制作方法 Method

（1）In a medium pot, heat oil over medium-high heat. Add onion and cook until translucent for about 5 minutes. Add zucchinis and mushrooms and cook until vegetables soften slightly for about 4 minutes. Add garlic and cook until fragrant for about 30 seconds. Add tomato puree, season with salt and black peppercorn, and bring mixture to a boil. Reduce heat to a rapid simmer and cook until zucchini is crisp-tender for about 8 minutes. Stir in oregano.

在锅中倒入油，中火加热，加入洋葱炒至半透明，约 5 min。加入西葫芦和蘑菇炒至蔬菜稍稍变软，约 4 min。再加入大蒜炒出香味，约 30 s。加入番茄酱，用盐和黑胡椒调味，煮至沸腾。将火调小至锅内微沸，煮至西葫芦口感脆嫩，大约 8 min，后加入牛至叶并搅拌。

（2）Meanwhile, in a large pot of boiling salted water, cook rigatoni until al dente. Reserve 1 cup of rigatoni water, drain rigatoni and add to sauce, toss to combine and add enough rigatoni water to create a sauce that coat rigatoni.

在一个大锅内将盐水煮沸,加入波纹贝壳状通心粉煮至口感稍硬。预留一杯煮波纹贝壳状通心粉的水,将波纹贝壳状通心粉沥干水后倒入酱汁内,搅拌均匀后加入煮波纹贝壳状通心粉的水直至酱汁的浓度完全裹上波纹贝壳状通心粉即可。

(四)奶油素食调味饭(creamy veggie risotto)

1 tablespoon of olive oil	1 汤匙橄榄油
1 onion, chopped	1 个洋葱,切碎
2 carrots, finely diced	2 根胡萝卜,切丁
300 g risotto rice	300 g 意大利米饭
1 bay leaf	1 片月桂叶
1 L vegetable stock	1 L 蔬菜汤
120 g frozen pea	120 g 冷冻青豆
50 g Parmesan cheese, grated	50 g 帕玛森奶酪,磨碎的

large shallow pan 大浅锅

(1) Heat the oil in a large shallow pan. Add the onion and carrots, cover and gently fry for about 8 minutes until the onion is very soft.

在一个大浅锅中加入油,后加入洋葱、胡萝卜,加盖后轻微炒制约 8 min 直到洋葱变软。

(2) Stir in the rice and bay leaf, then gently fry for another 2 minutes. Add 300 mL of the vegetable stock and simmer over a gentle heat, stir until it has all been absorbed. Add the hot vegetable stock, a ladleful at a time, let it be absorbed before adding more.

拌入米饭和月桂叶后再炒制 2 min,加入 300 mL 蔬菜汤后用小火炖,不停搅拌直至汤汁完全被吸收。分次加入蔬菜汤直至米饭煮熟,确保每次加入前汤汁完全被米饭吸收。

(3) Remove the bay leaf from the cooked risotto and stir in the peas. Heat through for a few minutes, then add most of the Parmesan cheese and season to taste. Sprinkle with the remaining Parmesan cheese and serve.

将月桂叶从米饭中取出,拌入青豆,煮制几分钟后,加入大部分的帕玛森奶酪后调味,最后撒上剩下的帕马森奶酪在饭的表面。

(五)奶油胡瓜千层面(creamy courgette lasagne)

9 dried lasagne sheets　　　　　　　　9 片干制意大利宽面条
1 tablespoon of sunflower oil　　　　　1 汤匙向日葵油
1 onion, finely chopped　　　　　　　1 个洋葱,切碎
5 courgette, coarsely grated　　　　　5 根小胡瓜,磨碎
2 garlic cloves, crushed　　　　　　　2 瓣大蒜,压碎
200 g ricotta　　　　　　　　　　　　200 g 意大利乳清干酪
50 g Cheddar　　　　　　　　　　　　50 g 切达干酪
300 g tomato sauce　　　　　　　　　300 g 番茄少司

saucepan　　　　　　　　　　　　　少司锅
frying pan　　　　　　　　　　　　　煎锅
baking dish　　　　　　　　　　　　烤盘

(1) Heat oven to 220 ℃. Bring a pan of water to the boil, then cook the lasagne sheets for about 5 minutes until softened, but not cooked through. Rinse in cold water, then drizzle with a little oil to stop them sticking together.

预热烤箱至 220 ℃,将一锅水煮沸后,加入意大利宽面条煮 5 min 直至变软,但不要完全煮熟,在冷水下冲洗面条,后撒上少量油防止其粘连。

(2) Meanwhile, heat the oil in a large frying pan, then fry the onion. After 3 minutes, add the courgettes and garlic and continue to fry until the courgette has softened and turned bright green. Stir in 2/3 of both the ricotta and the Cheddar, then season to taste. Heat the tomato sauce in the microwave until hot.

在煎锅中加入油后,炒香洋葱,约 3 min 后加入小胡瓜和大蒜继续炒至胡瓜变软、呈亮绿色,拌入 2/3 的意大利乳清干酪和切达干酪,调味。将番茄少司放入微波炉中加热。

(3) In a large baking dish, layer up the lasagne, start with half the courgette mix, then lasagne sheets, then tomato sauce. Repeat, top with blobs of the remaining ricotta, then scatter with the rest of the Cheddar. Bake about 10 minutes until the

lasagne sheets is tender and the cheese is golden.

在一个大的烤盘中开始制作千层面,依次放入一半的小胡瓜混合物、意大利宽面条、番茄少司,重复这个顺序,在最上面放剩下的意大利乳清干酪,撒上剩下的切达干酪。烤制约 10 min 直至意大利宽面条变软,奶酪呈金黄色即可。

第十章　西式甜点的制作工艺

西式甜点(dessert)是西餐饮食文化中不可或缺的一部分,通常在正餐后食用。英语"dessert"一词最早起源于法语"desservir",意思是从餐桌上撤去餐具,准备上甜点或水果,后来逐渐变成餐后甜点的意思。虽然"甜点"一词起源于法国,但是人类制作甜点却是从古埃及人发明蜂蜜圆饼开始的。

甜点的种类比较多,主要有蛋糕类、派类、冷冻类等。西式甜点的脂肪、蛋白质含量较高,味道香甜而不腻口,式样美观。西式甜点的取材多样,主要原料是面粉、糖、黄油、牛奶等,原材料的称量是否准确直接影响着最后成品的成败。

第一节　蛋糕的制作工艺

一、蛋糕概述

蛋糕是西点中最常见的品种之一。传说最早起源于中东地区,随后传入意大利、法国、英国等地,在17世纪后风靡欧洲各国,然后传入我国。蛋糕具有组织松软、香甜适口、装饰美观等特点,在其配料中鸡蛋、黄油的含量高,因而营养丰富。制作的蛋糕根据其使用原料、搅拌方法和面糊性质的不同,可分为海绵蛋糕、油脂蛋糕等。

二、海绵蛋糕(sponge cake)

海绵蛋糕是利用蛋白的起泡性使蛋液中充入大量的空气,再加入面粉烘烤而成的一类膨松点心,因为其结构类似于多孔的海绵而得名。在国外被称为泡沫蛋糕,在国内被称为清蛋糕。

在蛋糕的制作过程中,通过高速搅拌蛋白降低了其中的球蛋白的表面张力,增加了蛋白的黏度。因为黏度大的成分有助于初期泡沫的形成,所以能使之快速地打入空气,形成泡沫。蛋白中的球蛋白和其他蛋白,受搅拌的机械作用而产生了轻度变性。变性的蛋白质分子可以凝结成一层皮,形成十分牢固的薄膜将混入的空气包围起来。同时,由于表面张力的作用,蛋白泡沫收缩变成球形,加上蛋白胶体具有黏度,且加入的面粉原料附着在蛋白泡沫周围,从而使泡沫变得很稳定,能保持住混入的气体,在加热的过程中,泡沫内的气体又受热膨胀,使制品疏松多孔并具有一定的弹性和韧性。

制作海绵蛋糕的用料有鸡蛋、白糖、面粉及少量油脂等,其中新鲜的鸡蛋是制作海绵蛋糕最重要的条件,因为新鲜的鸡蛋胶体溶液的稠度高,能打进气体,并能保持气体性能稳定。存放时间长的鸡蛋不宜用来制作蛋糕。制作蛋糕的面粉多选择低筋粉,其粉质要细,面筋要软,但又要有足够的筋力来承担烘焙时的胀力,为形成蛋糕特有的组织起到骨架作用。制作蛋糕的糖多选择蔗糖,以颗粒细密、颜色洁白者为佳,如绵白糖或糖粉。颗粒大者,往往在搅拌时间短时不容易溶化,易导致蛋糕的质量下降。

由于海绵蛋糕在制作过程中所使用的鸡蛋成分不同,一种是只用蛋清而不用蛋黄的"天使蛋糕",另一种是用全蛋的"全蛋海绵蛋糕",因此两种蛋糕的配方各有不同。

海绵蛋糕的质量标准:表面呈金黄色,内部呈乳黄色,色泽均匀一致,糕体较轻,顶部平坦或略微凸起,组织细密均匀,无大气孔,柔软而有弹性,内无生心,口感不黏不干,轻微湿润,蛋味甜味相对适中。

(一) 天使蛋糕(angel cake)

天使蛋糕具有棉花般的质地和颜色,它是用硬性发泡的鸡蛋清、白糖和面粉制成的。由于不含牛油、油质,因此鸡蛋清的泡沫能更好地支撑蛋糕。天使蛋糕需要先将鸡蛋的蛋黄和蛋清分离,只用蛋清进行打发,整个配方中不要蛋黄,再配合面粉、白糖制作而成,因此它才会呈现白色。

制作天使蛋糕首先需将鸡蛋清打成硬性发泡(stiff peaks formed),然后用轻巧的翻折手法(folding)拌入其他材料。由于天使蛋糕不含油脂,因此口味和材质都非常的轻。天使蛋糕很难用刀切开,因此,通常需要使用叉子、锯齿形刀以及特殊的切具。

天使蛋糕需要使用专门的天使蛋糕烤具,通常是一个高身、圆筒状、中间有筒的容器,如图10-1(彩图23)所示。天使蛋糕烤好后,要倒置放凉以保持其体积。

(二) 全蛋海绵蛋糕(whole egg sponge cake)

全蛋海绵蛋糕与天使蛋糕的不同点在于其制作过程不仅使用了蛋清,同时也使用了蛋黄。

图10-1

1. 戚风蛋糕(chiffon cake)

戚风蛋糕属海绵蛋糕类型,制作原料主要有菜油、鸡蛋、糖、面粉、发粉等。但是由于其缺乏牛油蛋糕的浓郁香味,戚风蛋糕通常需要使用味道浓郁的汁,或加上巧克力、水果等配料。

由于菜油不像牛油那样容易打泡,因此需要靠把鸡蛋清打成泡沫状来提供足够的空气以支撑蛋糕的体积。戚风蛋糕含足量的菜油和鸡蛋,因此其质地非常的湿润,不像传统牛油蛋糕那样容易变硬。

先将鸡蛋的蛋黄和蛋清分离,然后将蛋黄与面粉混合成蛋黄糊,再将蛋白加糖单独打发,之后将二者混合拌匀,烤制而成戚风蛋糕。

戚风蛋糕的组织蓬松,水分含量高,味道清淡不腻,口感滋润嫩爽。若将相同重量的全蛋搅拌式海绵蛋糕面糊与戚风蛋糕的面糊同时烘烤,则戚风蛋糕的体积可能是前者的两倍。

2. 清蛋糕胚

制作天使蛋糕只使用蛋白,制作戚风蛋糕要先将蛋黄和蛋清分开,待蛋清打发后再一起拌匀。而海绵蛋糕是将整个鸡蛋放入打蛋盆中,全蛋打发,再配合油脂和粉类、白糖制作而成的。

(三)海绵蛋糕制作实例

1. 天使蛋糕制作实例

  原料　Ingredients

200 g egg white	200 g 蛋白
140 g white sugar	140 g 白砂糖
65 g cake flour, sifted	65 g 筛过的低筋面粉
10 g corn starch	10 g 玉米淀粉
tartar powder, a bit	塔塔粉,少量
1 g salt	1 g 盐

  工具　Tools

bowl	搅拌碗
wooden spatula	木铲
whisk	搅拌器

  制作方法　Method

(1) Add salt, tartar powder to the egg white, whisk the egg white, add sugar in 3 times, continue whisking egg white until it forms stiff peaks.

将盐、塔塔粉加入蛋白中,后分三次加入白砂糖至蛋白中,搅打蛋白至起泡。

(2) Add sifted cake flour, corn starch to the whipped egg white.

将低筋面粉和玉米淀粉混合过筛后,倒入蛋白中。

(3) Gently combine the egg white with the dry ingredients, and then pour into the mould.

将面粉混合物与蛋白搅拌均匀,后倒入 8 寸(26.67 cm)模具中。

(4) Smooth the batter and shake the mould, and bake for about 40 minutes at 190 ℃ in the oven.

抹平蛋糕表面,晃动模具,以便把内部的大气泡排出,放入 190 ℃ 的烤箱烤制约 40 min。

(5) Invert cake, and allow it to cool.

倒出蛋糕,放凉。

2. 戚风蛋糕制作实例

 原料 Ingredients

5 eggs, separated	5 个鸡蛋(蛋清、蛋白分开)
85 g cake flour, sifted	85 g 筛过的低筋面粉
40 g vegetable oil	40 g 色拉油
40 g milk	40 g 牛奶
60 g white sugar(for egg white)	60 g 白砂糖(加入蛋白中)
30 g white sugar(for egg yolk)	30 g 白砂糖(加入蛋黄中)

 工具 Tools

bowl	搅拌碗
wooden spatula	木铲
whisk	搅拌器

 制作方法 Method

(1) Prepare all the ingredients. Separate the egg white and egg yolk to the bowl.

准备好所有材料,将蛋清、蛋黄分离到无水无油的盆中。

(2) In a large mixing bowl, beat egg white with sugar until stiff. Set aside.

用打蛋器把蛋白打发至呈鱼眼泡状时,加入 20 g 的白砂糖,继续搅打至蛋白开始变浓稠,再加入 20 g 糖,继续搅打,到蛋白表面出现纹路的时候,最后加入剩下的 20 g 糖。当搅打时感觉有阻力,则停止打发。

(3) Add 30 g sugar to the egg yolk, whisk together gently.

加入 30 g 白砂糖到蛋黄中,用打蛋器将其轻轻打散,不要用力搅拌,以免打发蛋黄。

(4) Add 40 g vegetable oil, 40 g milk to the egg yolk mixture, mix well, and then add sifted flour, fold well. Do not over mix.

依次加入 40 g 色拉油、40 g 牛奶到蛋黄、糖的混合物中,搅拌均匀,后加入面粉,翻拌均匀。不要过度搅拌,以免面粉起劲。

(5) Add 1/3 whisked egg white to the egg yolk mixture, fold well. Then add all the egg yolk to the bowl with leftover egg white mixture, fold well.

取 1/3 的打发蛋白加入蛋黄混合物中,翻拌均匀。后将蛋黄糊全部倒入装有剩下的蛋白的容器中,直至蛋白和蛋黄糊完全混合。

(6) Pour the cake batter to the mould, shake the mould, and bake for about 1 hour at 170 ℃ in the oven. Invert pan until cool.

将混合好的蛋糕糊倒入 8 寸(26.67 cm)的圆形模具中,晃动模具,以便把内部的大气泡排出。将装有蛋糕糊的模具放入 170 ℃ 的烤箱烤约 1 h。将烤好后的蛋糕立即倒扣直到冷却,最后脱模。

三、油脂蛋糕

油脂蛋糕是使用鸡蛋、黄油、面粉、糖等原料进行搅拌、烘焙制成的。其含有较多的固体油脂,弹性和柔软度不及海绵蛋糕,但其结构组织紧密,口感细腻,具有较强的饱腹感。

(一)制作原理

1. 打发油脂

在蛋糕的制作过程中,搅拌作用使空气进入油脂形成气泡,从而使油脂膨松、体积增大。油脂越蓬松越好,蛋糕质地则会越疏松,但蓬松过度会影响蛋糕的成型。

2. 油脂与蛋液的乳化

将蛋液加入打发的油脂中时,蛋液中的水分与油脂在搅拌下发生乳化。乳化对油脂蛋糕的品质有重要影响,乳化越充分,蛋糕制品的组织越均匀,口感越佳。

油脂蛋糕乳化时容易发生油、水分离的现象,其原因主要有以下几点。

(1)所用油脂的乳化性差。

(2)浆料的温度过高或过低(最佳温度为 21 ℃)。

(3)蛋液加得太快,每次未充分搅拌均匀。

为了改善油脂的乳化,在加蛋液的同时可加入适量的蛋糕油(为面粉量的 3%～5%)。蛋糕油作为乳化剂,可使油和水形成稳定的乳液,使蛋糕的质地更加细腻,并能防止产品老化,延长其保鲜期。

(二)油脂蛋糕制作实例

1. 玛德琳蛋糕(Madeleine cake)

玛德琳蛋糕是一种法国风味的小甜点,又叫贝壳蛋糕。由黄油、低筋面粉等为主料,及糖、泡打粉、全蛋等配料组合制作而成。

 原料 Ingredients

80 g cake flour, sifted	80 g 低筋面粉,过筛
80 g egg	80 g 鸡蛋液
80 g butter, melted	80 g 融化的黄油
80 g white sugar	80 g 白砂糖
1/2 teaspoon of baking powder	半茶匙泡打粉
1/4 teaspoon of vanilla extract	1/4 茶匙香草精
zest of 1/2 lemon	半个柠檬的皮

 工具 Tools

mixing bowl	搅拌碗
wooden spatula	木铲
whisk	搅拌器
grater	磨皮器

 制作方法 Method

(1) Mix lemon zest and white sugar in the mixing bowl.

在一个大碗里混合柠檬皮和白砂糖。

(2) Beat eggs in a bowl, then add the mixture of sugar and lemon zest, vanilla extract, mix well all the ingredients in the bowl.

将鸡蛋打入碗中,后加入糖、柠檬皮的混合物,并加入香草精华,搅拌均匀。

(3) Add sifted cake flour, baking powder to the egg mixture, mix well, then add melted butter, mix.

加入筛过的面粉、泡打粉至鸡蛋混合物中,搅拌均匀,最后加入融化的黄油。

(4) Cover the batter with plastic film and refrigerate for 1 hour.

用保鲜膜盖住面糊,放入冰箱内冷藏 1 h。

(5) Grease the mould, squeeze the batter into the mould.

将面糊倒入刷油后的模具。

(6) Bake for 15 minutes at 190 ℃ in the oven.

将装有面糊的模具放入190 ℃的烤箱中烤制约15 min。

2. 蝴蝶纸杯蛋糕(butterfly cupcake)

杯型蛋糕又称口袋蛋糕,在英国通常被称为fairy cake,即仙女蛋糕,在美国也被称为patty cake or cup cake。

 原料 Ingredients

60 g unsalted butter, softened	60 g 无盐黄油,软化
80 g caster sugar	80 g 细白砂糖
150 g cake flour, sifted	150 g 低筋面粉,过筛
60 mL milk	60 mL 牛奶
1 egg	1个鸡蛋
60 mL cream	60 mL 奶油
40 g strawberry jam	40 g 草莓果酱
icing sugar, to dust	糖粉

 工具 Tools

mixing bowl	搅拌碗
electric beater	电动搅拌器

 制作方法 Method

(1) Preheat the oven to 180 ℃.

预热烤箱至180 ℃。

(2) Place the unsalted butter, sugar, flour, milk and eggs in a large mixing bowl. Using the electric beater, beat on low speed and then increase the speed until the mixture is smooth and pale. Divide the mixture evenly among the cases and bake for 30 minutes, transfer to a wire rack to cool.

将无盐黄油、糖、面粉、牛奶、鸡蛋混合在大碗里,使用电动搅拌器搅打混合物,先低速搅打,后加速搅打直至混合物变得光滑,颜色变淡。将混合物分至模具中烤制30 min,后放在铁架上冷却。

(3) Cut shallow round from the centre of each cake using the point of a sharp knife, then cut the round in half. Spoon 2 teaspoons of cream into each cavity, top with 1

teaspoon of jam and position two halves of the cake tops in the jam to resemble butterfly wings. Dust with icing sugar.

从蛋糕的中间切开一个浅圆形,然后将这个浅圆形切成两半。在切开的洞中放入 2 茶匙奶油,然后顶部放入 1 茶匙草莓酱,放上刚切下来的蛋糕作为蝴蝶的翅膀,最后撒上糖粉。

3. 黑巧克力蛋糕(rich dark chocolate cake)

原料　Ingredients

185 g unsalted butter, softened	185 g 无盐黄油,软化
250 g dark chocolate chips	250 g 黑巧克力碎
220 g cake flour, sifted	220 g 低筋面粉,过筛
40 g unsweetened cocoa powder	40 g 无糖可可粉
375 g caster sugar	375 g 细白砂糖
3 eggs, lightly beaten	3 个鸡蛋,打散
chocolate topping	巧克力黄油酱
20 g unsalted butter, chopped	20 g 无盐黄油,切碎的
125 g dark chocolate, chopped	125 g 黑巧克力,切碎的

工具　Tools

spring-form cake mould	弹簧形蛋糕模具
mixing bowl	搅拌碗
saucepan	少司锅

制作方法　Method

(1) Preheat oven to 160 ℃, lightly grease the spring-form cake mould and line the base with baking paper.

预热烤箱至 160 ℃,将弹簧形蛋糕模具刷油后铺上烤纸。

(2) Place the butter and chocolate chips in a small heatproof bowl. Place the bowl over a saucepan of simmering water, make sure the base does not touch the water, and stir frequently until melted.

将黄油、巧克力碎放入耐热的碗里,将碗隔水加热,不断搅拌直至黄油、巧克力融化。

(3) Sift the flour and cocoa powder into a large mixing bowl. Combine the chocolate mixture, sugar and eggs, then add 250 mL water and mix well. Add to the flour and cocoa powder and stir until well combined. Pour the mixture into the prepared mould

and bake for 1.5 hours, Then leave the cake in the mould for 15 minutes before turning out onto a wire rack to cool completely.

将面粉、可可粉筛入大碗中。混合巧克力-黄油混合物、糖、鸡蛋,后倒入 250 mL 水,待其混合均匀后加入面粉混合物中搅拌均匀。最后将混合物倒入准备好的模具中烤制 1.5 h,静置 15 min 后脱模冷却。

(4) To make the topping, place the butter and chocolate chips in a small heatproof bowl. Place the bowl over a saucepan of simmering water, make sure the base does not touch the water. Spread over the cooled cake in a swirl pattern.

制作巧克力黄油酱时,先将黄油、巧克力碎放入耐热的碗里,再将碗隔水加热搅拌直至黄油、巧克力融化。最后将巧克力黄油酱旋流形淋在蛋糕上。

4. 巧克力玛芬蛋糕(chocolate muffin)

玛芬蛋糕(又称松饼),有多种口味,是西方很受欢迎的一种点心。

 原料 Ingredients

85 g low gluten flour	85 g 低筋面粉
2 tablespoons of cocoa powder	2 汤匙可可粉
60 g butter	60 g 黄油
85 g icing sugar	85 g 糖粉
1 egg	1 个鸡蛋
80 mL milk	80 mL 牛奶
1/8 teaspoon of salt	1/8 茶匙盐
1/2 teaspoon of baking powder	1/2 茶匙泡打粉
1/4 teaspoon of baking soda	1/4 茶匙小苏打

 工具 Tools

whisk　　　　　　　　　　　　　　　搅拌器
rubber spatula　　　　　　　　　　　橡皮刮刀

 制作方法 Method

(1) Sift low gluten flour, cocoa powder, baking powder and baking soda.

把低筋面粉、可可粉、泡打粉、小苏打混合过筛备用。

(2) Soft the butter, whisk them (until they have smooth surface, pale colour, bigger volume), add icing sugar to the butter and whisk together.

软化黄油后,用打蛋器稍微打发(直至混合物的表面光滑,颜色发白,体积变大),并加入糖粉搅匀。

(3) Add egg liquid in three times and mix well, make sure the egg and butter are thoroughly blended before adding egg in each time(the mixture should be exquisite and smooth, the egg and oil can not be separated).

分三次加入打散的鸡蛋液,并搅拌均匀。每一次都需要使鸡蛋和黄油完全融合后再加下一次(混合物应细腻光滑,不出现蛋油分离)。

(4) Pour the milk to the mixture(do not stir).

倒入牛奶(不需要搅拌)。

(5) Add the sifted flour mixture.

加入第一步中筛过的面粉混合物。

(6) Fold the mixture with rubber spatula until smooth.

用橡皮刮刀轻轻翻拌均匀,至混合物光滑无颗粒。

(7) Pour the batter to the paper cup, do not fill it to the top.

将面糊倒入纸杯,不要将纸杯装满。

(8) Bake for about 30 minutes at 180 ℃ in the oven.

放入 180 ℃ 的烤箱中烤制约 30 min。

四、奶酪蛋糕

奶酪蛋糕是指用奶酪作为主要原料制作的蛋糕,是西式甜点的一种。奶酪蛋糕起源于古老的希腊,是在公元前 776 年为了供应雅典奥运所做出来的甜点,接着由罗马人将奶酪蛋糕从希腊传播到整个欧洲,又在 19 世纪随着欧洲移民们传到了美洲。奶酪蛋糕有着柔软的上层,混合了特殊的奶酪,再加上糖和其他的配料,如鸡蛋、奶油、水果等。奶酪蛋糕通常都以饼干作为底层,此类蛋糕在结构上较一般的蛋糕扎实,但质地却比一般蛋糕更绵软,口感湿润。

(一) 奶酪蛋糕(cheese cake)

250 g cream cheese　　　　　　　　250 g 奶油奶酪

80 g caster sugar　　　　　　　　　80 g 白砂糖

2 eggs　　　　　　　　　　　　　　2 个鸡蛋

15 g corn starch　　　　　　　　　　15 g 玉米淀粉

10 g lemon juice　　　　　　　　　　10 g 柠檬汁

80 g milk	80 g 牛奶
1 tablespoon of rum	1 汤匙朗姆酒
1/4 teaspoon of vanilla extract	1/4 茶匙香草精华
cake base	蛋糕底
100 g digestive biscuit crumbs	100 g 消化饼干碎
50 g melted butter	50 g 黄油,融化的

 工具 Tools

6-inch round mould	6寸(20 cm)圆形模具
spoon	勺子
whisk	搅拌器
roasting pan	烤盘

 制作方法 Method

(1) Add digestive biscuit crumbs to the melted butter, mix together, put the mixture to the mould, press down on the crumbs with spoon until the crumbs become a nice even layer at the bottom of the mould, then refrigerate.

将消化饼干碎末倒入黄油里,搅拌均匀后铺在蛋糕模底部,用小勺压平压紧。铺好蛋糕底后,把蛋糕模放进冰箱冷藏备用。

(2) Soft the cream cheese at the room temperature, then add sugar, whisk together until smooth.

将奶油奶酪放在室温下软化,加入细砂糖,用打蛋器打至顺滑、无颗粒的状态。

(3) Add eggs in batches, whisk the mixture until well mixed.

分次加入鸡蛋,并用打蛋器搅打均匀。

(4) Pour lemon juice, whisk together the mixture.

倒入柠檬汁,搅打均匀。

(5) Stir in the corn starch, milk, rum, vanilla extract and mix well.

倒入玉米淀粉、牛奶、朗姆酒、香草精,搅打均匀。

(6) Pour the batter to the mould.

把蛋糕糊倒入铺好蛋糕底的蛋糕模里。

(7) Carefully pour the hot water into the roasting pan, to create a water bath for the cheesecake, pouring until the water reaches halfway up the side of the mould.

将蛋糕模放入烤盘,在烤盘里倒入热水,热水高度高于模具高度的一半即可。

(8) Put the roasting pan in the oven and bake for 1 hour at 160 ℃ until the colour is golden.

把烤盘放入预热至 160 ℃ 的烤箱,烤 1 h,蛋糕表面呈金黄色即可出炉。

(9) Remove from the oven and refrigerate for 4 hours.

将烤好的蛋糕冷藏 4 h 后,脱模并切块食用。

(二) 蔓越莓奶酪蛋糕(cranberry cheese cupcake)

 原料 Ingredients

120 g digestive biscuit	120 g 消化饼干
50 g butter, melted	50 g 融化的黄油
300 g cream cheese	300 g 奶油芝士
1 tablespoon of cake plain flour	1 汤匙低筋面粉
80 g caster sugar	80 g 白砂糖
dash vanilla extract	少许香草精华
1 egg, plus 1 yolk	1 个鸡蛋,加上 1 个蛋黄
200 g cranberries	200 g 蔓越莓
50 g icing sugar	50 g 糖粉

 工具 Tools

muffin tin	松饼模
muffin case	松饼纸杯
spoon	勺子

 制作方法 Method

(1) Put the cranberries and icing sugar into a shallow pan and cook over medium heat for 10 minutes until the mixture become thick and sticky sauce. Turn off the heat and leave to cool.

在一个浅锅中混合蔓越莓、糖粉,用中火将蔓越莓和糖粉熬制 10 min 直至呈黏稠少司状,后关火,冷却混合物。

(2) Heat oven to 180 ℃. Line a 6-hole muffin tin with 6 muffin cases. Crush the biscuits in a plastic bag, then mix with the melted butter. Divide between the muffin cases and press down with your fingers.

预热烤箱至 180 ℃,将 6 孔松饼模分别套上松饼纸杯,将饼干放入塑料袋中捣碎后与黄油混合。将混合物倒入纸杯中,后用手指按压。

(3) Mix the soft cheese with the flour, sugar and vanilla extract in a bowl, then gradually beat in the eggs and yolk until smooth. Ripple the cranberry mixture through

the cheese, do not over-mix. Spoon the mixture into the cases and smooth the tops with the back of the spoon. Bake for 30 minutes, leave to cool, then chill in the fridge.

在碗中混合奶酪、面粉、糖、香草精华,然后缓慢加入鸡蛋和蛋黄直至混合物变得光滑。混合蔓越莓混合物与奶酪混合物,不要过度搅拌。用勺子将混合物装入纸杯中,后用勺子的背面将混合物表面抹平。烤制 30 min 后冷却,后放入冰箱冷藏。

第二节　派类制作工艺

一、派的分类

派是英文"pie"的译音,派是由派馅及派皮两部分烤制而成的一种甜点,在各种西点中具有独特的风味。一个上好的派需要有调制适宜的派馅以及酥软松脆的派皮。根据派馅及制作程序的不同,一般可分为单层皮派、双层皮派两大类,按口味可分为甜、咸两种。

(一) 单层皮派

单层皮派是由一层派皮上面盛装各种馅料而制成的,如南瓜派、柠檬布丁派、香蕉派等。

(二) 双层皮派

双层皮派是用两片派皮将煮好的馅包在中间,然后进炉烘烤。它又分为以下几种。
(1) 水果派,使用较硬的水果做馅,如苹果派、菠萝派等。
(2) 肉派,使用牛肉、鸡肉等作为馅料。
(3) 油炸派,如油炸苹果派、樱桃派等。

二、派类制作工艺

(一) 派皮的制作(pie crust)

 原料　Ingredients

100 g low gluten flour　　　　　　　　100 g 低筋面粉

40 g butter   40 g 黄油

10 g caster sugar   10 g 细砂糖

33 g water   33 g 水

mixing bowl   搅拌碗

rolling pin   擀面棍

(1) Soften butter, add flour, sugar, rub butter and flour constantly until well mixed.

软化黄油,后倒入低筋面粉和细砂糖,不断揉搓黄油和面粉,直至搓匀。

(2) Add water to the mixture to form dough, then stand for about 15 minutes.

在混合物中加入水,揉成面团后,静置约 15 min。

(3) Roll out the pastry.

将静置好的面团擀成薄片。

(4) Cover the pie plate with the pastry.

将薄片盖在派盘上。

(5) Gently press the pastry down so that it lines the bottom and sides of the pie plate.

用手轻压面片,使其和派盘贴合紧密。

(6) Trim the pastry.

移去多余的面片。

(二)苹果派(apple pie)

225 g plain flour   225 g 中筋面粉

140 g butter   140 g 黄油

3 large apples   3 个大苹果

2 tablespoons of honey   2 汤匙蜂蜜

pinch of cinnamon   少量肉桂粉

1 egg   1 个鸡蛋

vanilla ice cream, to serve   香草冰激凌

工具 Tools

mixing bowl	搅拌碗
rolling pin	擀面棍
pie dish	大盘

 制作方法 Method

(1) Heat oven to 200 ℃. To make the pastry, sift the flour into a large mixing bowl and add the butter, mix well.

预热烤箱至 200 ℃。制作面皮，在大碗里筛入面粉，加入黄油，充分混合。

(2) Add 3 tablespoons of cold water(1 tablespoon at a time) to the mixture. Then wrap it in the cling film and leave to chill in the fridge for 30 minutes.

加入 3 汤匙冷水至面粉混合物中，分 3 次加入，每次 1 汤匙，后用保鲜膜包裹面团，放入冰箱中冷藏 30 min。

(3) While the pastry is chilling, core the apples, then cut them into even-sized chunks. Put the apples into the pie dish, drizzle over the honey, add the cinnamon and 2 tablespoons of water.

去除苹果的果核，将其切成大小一致的块，将切块苹果放入大盘中，滴入蜂蜜，加入肉桂和 2 汤匙水。

(4) Roll out the pastry until it is large enough to cover the pie dish, carefully lift the pastry and lay it over the top of the apple mixture. Carefully trim off the excess pastry and press the pastry edges onto the dish to create a seal.

将面团擀薄直至能够平铺大盘，小心地将面皮铺在苹果上方，并去掉多余的面皮。用手按压盘子边缘的面皮，使其密封。

(5) Make a small cut in the pastry so that the air can escape during cooking, then brush with beaten egg to increase luster.

在面皮上打些小孔（烤制时有利于空气的排出），然后刷上蛋液以增加光泽。

(6) Bake the pie in the oven for 20-30 minutes until the pastry is golden and the apple filling is bubbling and hot. Serve when it is still warm with ice cream.

在烤箱中烤制派 20～30 min 直至其面皮呈金黄色，苹果馅料冒气泡。可搭配香草冰激凌。

(三)饼干青柠派(biscuity lime pie)

 原料 Ingredients

300 g ginger nut biscuit	300 g 生姜坚果饼干
100 g butter, melted	100 g 融化的
3 egg yolks	3 个蛋黄
50 g caster sugar	50 g 细砂糖
1 lime	1 个青柠
1 lemon	1 个柠檬
300 g sweetened condensed milk	300 g 甜炼乳

 工具 Tools

food processor	搅拌器
mixing bowl	搅拌碗

 制作方法 Method

(1) Heat oven to 180 ℃. Tip the biscuits into a food processor and blitz to crumbs. Add the butter to combine. Tip the mixture into a fluted rectangular tart tin, about 10 cm×34 cm and press them into the base and up the sides right to the edge. Bake for 15 minutes until crisp.

预热烤箱至 180 ℃。把饼干放入搅拌器中迅速打成面包屑状,后加入黄油,将混合物放入有凹槽的方形模具中(约为 10 cm×34 cm),按压混合物,将其四周对齐,烤制 15 min 直至焦脆。

(2) While the base is baking, tip the egg yolks, sugar, lime and lemon zests into a bowl and beat with an electric whisk until doubled in volume. Pour in the condensed milk, beat until combined, then add the lemon juice.

将蛋黄、糖、青柠皮、柠檬皮放入碗中,并用电动搅拌器搅拌直至体积翻倍,倒入炼乳,搅拌至完全混合,最后加入柠檬汁。

(3) Pour the mixture into the tart case and bake for 20 minutes until just set with a slight wobble in the centre. Leave to set completely, then remove from the tin, cool and chill. Serve in slices topped with thin lime slices, if you like.

将混合物倒入模具中,烤制约 20 min(初步定型,中间部分略有晃动)。静置后脱模,放凉冷藏。切片后用青柠片装饰顶部。

第三节　泡芙类制作工艺

泡芙(puff)是一种源自意大利的甜食。它的蓬松发胀的奶油面皮中包裹着奶油、巧克力、冰激凌。泡芙吃起来外热内冷，外酥内滑，口感极佳。

Puff pastry is made by layering dough with butter, and it is folded to creat hundreds of layers. When cooked, the butter melts and the dough produces steam, forcing the layers apart and making the pastry rise to great height.

泡芙面团是将面团折叠分层，当烘焙时，黄油融化、面团产生蒸汽，从而形成较强的蒸汽压力，将面皮撑开，形成泡芙。

泡芙面团(puff pastry)的制作如下。

 原料　Ingredients

100 g cake flour, sifted	100 g 低筋面粉，过筛
160 g water	160 g 水
80 g butter	80 g 黄油
1 teaspoon of sugar	1 茶匙糖
1/2 teaspoon of salt	半茶匙盐
3 eggs	3 个鸡蛋

 工具　Tools

saucepan	少司锅
wooden spatula	木铲
baking tray	烤盘

 制作方法　Method

(1) Add water, salt, sugar, butter to the saucepan, mix a bit over medium heat, lower the heat until boiling, then add all the flour.

将水、盐、糖、黄油一起放入锅里，用中火加热并略微搅拌，使黄油分布均匀。当煮至沸腾的时候，调至小火，然后一次性加入面粉。

(2) Stir the mixture quickly with wooden spatula until flour and water are well

mixed, then remove from the heat.

用木勺快速搅拌混合物,使面粉和水完全混合在一起,直至混合物充分融合后,关火。

(3) Start to add eggs when the temperature of batter lower to 60 ℃, when batter and eggs are well mixed add more eggs.

待面糊冷却到 60 ℃时,分次加入鸡蛋。

(4) Don't need to add eggs if the batter likes triangle and has 4 cm height when lifted.

挑起面糊,若面糊呈倒三角形状,尖角到底部约为 4 cm,并且不会滑落,这时不用再继续加入鸡蛋。

(5) Spoon the batter to the baking tray.

用小勺直接挖起泡芙面糊放在烤盘上(烤盘里垫上锡纸)。

(6) Preheat oven to 210 ℃, bake for 10-15 minutes, when the puff swell, lower the heat to 180 ℃ and bake for 20-30 minutes.

预热烤箱至 210 ℃,烤制 10~15 min。当泡芙膨胀起来以后,把温度降低到 180 ℃,继续烤 20~30 min。

(7) Cool puffs and stuff them.

待泡芙完全冷却后,填入馅料。

第四节 冷冻类甜品的制作工艺

一、冷冻类甜品的制作工艺

(一) 提拉米苏(Tiramisu)

Tiramisu is a classic Italian dessert. It is made of lady finger biscuits dipped in espresso or strong coffee, layered with a whipped mixture of egg yolks, Mascarpone cheese, sugar, and topped with cocoa.

提拉米苏是意大利著名的甜品,它以马斯卡彭奶酪作为主要材料,再以手指饼干取代海绵蛋糕,并加入了咖啡、可可粉等其他原料。吃到嘴里香、滑、甜、腻,柔和中带有质感。

手指饼干(lady finger)是意大利著名的饼干,它的外形细长,类似于手指的形状,质地干燥,非常香甜。由于其质地类似于干燥过的海绵蛋糕,能够吸收大量的水分,因此非常适合用来做提拉米苏的基底及夹层。手指饼干的制作如下。

 原料 Ingredients

1 egg yolk, 10 g white sugar　　　　　1个蛋黄,10 g 白糖
1 egg white, 15 g white sugar　　　　　1个蛋白,15 g 白糖
35 g low gluten flour　　　　　　　　35 g 低筋面粉

 工具 Tools

bowl　　　　　　　　　　　　　　搅拌碗
rubber spatula　　　　　　　　　　橡胶刮刀
baking pan　　　　　　　　　　　　烤盘

 制作方法 Method

(1) Separate egg yolk and egg white.
分开蛋黄、蛋白。
(2) Whip egg white, add sugar in three times.
打发蛋白,分三次加入白糖。
(3) Whip egg yolk with sugar.
打发蛋黄和糖。
(4) Add egg white mixture to the egg yolk mixture, slightly mix and fold.
将蛋白混合物加入蛋黄混合物中,略微翻拌。
(5) Sift flour to the mixture, slightly mix and fold.
筛入面粉至混合物中,轻轻混合翻拌。
(6) Put the mixture to the nozzle and make strips on the baking tray.
将混合物放入裱花嘴中,在烤盘中挤成长条状。
(7) Bake at 190 ℃ for 10 minutes.
放入 190 ℃烤箱中烤制 10 min。
提拉米苏的制作如下。

 原料 Ingredients

1 egg yolk　　　　　　　　　　　　1个蛋黄
120 g cream, whipped　　　　　　　120 g 奶油,打发

60 g sugar   60 g 白糖

25 mL strong espresso, 25 mL white rum wine

                                                        25 mL 意大利浓缩咖啡, 25 mL 白朗姆酒

250 g Mascarpone cheese   250 g 马斯卡彭奶酪

lady fingers   手指饼干

cocoa powder   可可粉

icing sugar   糖粉

工具 Tools

bowl   搅拌碗

rubber spatula   橡胶刮刀

制作方法 Method

（1）Place a saucepan 1/3 full of water onto the stove and bring to a simmer. Meanwhile, mix together the egg yolk and sugar in a medium bowl. Briefly whip them before placing the bowl over the simmering water. Continue to whip for a few minutes until the egg is slightly cooked, and it has a smooth, runny consistency.

在少司锅中倒入1/3的水煮沸，同时在碗里混合蛋黄和糖，简单地将它们混合后隔水搅拌直至混合物变得光滑。后离火，冷却。

（2）Next, spoon the Mascarpone cheese into a large bowl with your rubber spatula. Add the lightly cooked egg yolk into the Mascarpone cheese. Mix together until the cheese is fully incorporated and lump free. Now slowly spoon in the whipped cream and combine it well with your whisk.

用橡胶刮刀挖一勺马斯卡彭奶酪至大碗中，后慢慢加入蛋黄混合物，搅拌均匀直至奶酪完全融合，最后慢慢加入打发的奶油，搅拌均匀。

（3）Build the Tiramisu. Start by combining the espresso and the white rum wine together with a large spoon. Now dip the entire body of lady fingers into the mixture. Let them soak and absorb the mixture for a few seconds. Then place 2 lady fingers each horizontally into each cup. Next, spoon over the cream mixture to half fill the cups. Again, dip the whole body of the lady fingers into the espresso, mix and place three lady fingers each horizontally into each cup, on top of the cream.

混合意大利浓缩咖啡和白朗姆酒，将手指饼干完全浸入混合物中，浸泡吸收几秒钟。在杯中水平放置2根手指饼干，后铺上奶油混合物直至杯子容积的一半，接着水平放置3根手指饼干，顶部铺上奶油混合物。

（4）Place it into the fridge for a few hours to set. Before serving, dust with a little cocoa powder.

放入冰箱冷藏数小时定型,撒上少量可可粉装饰。

(二)焦糖布丁(cream caramel)

焦糖布丁是一道法国的经典甜品,主要是利用焦糖反应在成品的上方覆盖一层薄而脆的糖衣。焦糖反应是指糖类在高温(一般为150~200 ℃)加热的条件下发生降解,其降解产物经缩合、聚合形成了具有黏稠状特性的黑褐色物质。在制作焦糖的过程中一定要严格把握好火候,否则很容易烧成焦黑色,从而影响成品的质量。

  原料 Ingredients

90 g sugar	90 g 白糖
For custard	蛋奶糊
350 mL milk	350 mL 牛奶
40 g caster sugar	40 g 细砂糖
2 eggs	2 个鸡蛋
1/2 teaspoon of vanilla extract	半茶匙香草精华

  工具 Tools

saucepan	少司锅
whisk	搅拌器
baking dish	烤盘

  制作方法 Method

(1) Preheat oven to 160 ℃, brush mould with melted butter.
预热烤箱至160 ℃,将模具刷上融化的黄油。

(2) Place the sugar and 30 mL water in a saucepan. Stir over low heat until the sugar dissolves. Bring to the boil, reduce the heat and simmer, without stirring, until the mixture turns golden and starts to caramelise. Remove from the heat immediately and pour enough hot caramel into each ramekin to cover the base.
在少司锅中加入糖和30 mL水,小火搅拌直至糖完全融化。煮沸后调成小火炖直至混合物变成金黄色。当混合物开始变成焦糖后,立即离火,将足够的热焦糖倒入模具中。

(3) To make the custard, heat the milk in a pan over low heat until almost boiling. Remove from the heat. Whisk together the sugar, eggs and vanilla for about 2 minutes,

then stir in the warm milk. Strain the mixture into a jug and pour into the ramekins.

制作蛋奶糊时,先在锅中用小火加热牛奶直至几乎沸腾,离火。搅拌糖、鸡蛋、香草精华约 2 min,然后加入热牛奶,过滤混合物至罐子中,后将混合物倒入模具中。

（4）Place the ramekins in a baking dish and pour in enough boiling water to come halfway up the sides of the ramekins. Bake for 30 minutes. The custard should no longer be liquid and should wobble slightly when the dish is shaken lightly. Allow to cool, then refrigerate for at least 2 hours.

将模具放入烤盘中,倒入足够的开水至模具的半高位置,烤 30 min,烤至蛋奶糊不再是流体状,当晃动烤盘时,蛋奶糊会轻微地颤动即可。冷却后,冷藏至少 2 h。

（5）To unmould, run a knife carefully around the edge of each custard and gently upturn it onto the serving plates.

脱模时,先沿着蛋奶糊的边缘轻轻滑动刀,再轻轻地将其翻转到盘中。

（三）巧克力慕斯（chocolate mousse）

慕斯的英文是"mousse",它是一种奶冻式的甜点,可以直接吃或用来做蛋糕夹层,通常是加入奶油与凝固剂来达到呈浓稠冻状的效果。慕斯是从法语音译过来的,慕斯蛋糕最早出现在美食之都——法国巴黎,大师们在奶油中加入起稳定作用和改善结构、口感、风味的各种辅料,待其冷冻后食用。慕斯与布丁一样属于甜点的一种,其质地较布丁更为柔软,入口即化。

 原料 Ingredients

130 g dark chocolate, finely chopped	130 g 黑巧克力,切碎
30 g unsalted butter, diced	30 g 无盐黄油,切块
2 tablespoons of espresso	2 汤匙浓缩咖啡
150 mL cream	150 mL 奶油
3 large eggs, separated	3 个鸡蛋（蛋清、蛋黄分开）
1 tablespoon of sugar	1 汤匙糖

 工具 Tools

double boiler	双层蒸锅
rubber spatula	橡胶刮刀
whisk	搅拌器

制作方法 Method

(1) Whip the cream to form soft peaks, then refrigerate it.

打发淡奶油,冷藏。

(2) Combine the chocolate, butter and espresso in the top of a double boiler over hot, but not simmering water, stirring frequently until smooth. Remove the mixture from the heat and let it cool until the chocolate is just slightly warmer than body temperature.

混合巧克力、黄油、浓缩咖啡,后将其放置于双层蒸锅的上层,加热搅拌混合物直至光滑,后离火冷却直至巧克力的温度略高于人体温度。

(3) Add the egg whites. Once the melted chocolate has cooled slightly, whip the egg whites in a medium bowl until they are foamy and beginning to hold a shape. Sprinkle in the sugar and beat until soft peaks form.

当巧克力开始冷却时,在小碗中打发蛋白(其间加入糖)。

(4) Add the egg yolks. When the chocolate has reached the proper temperature, stir in the yolks. Gently stir in about one-third of the whipped cream. Fold in half the whites just until incorporated, then fold in the remaining whites, and finally the remaining whipped cream.

当巧克力达到预期温度时,加入蛋黄搅拌,后缓慢加入1/3的打发淡奶油,加入一半的打发的蛋白,翻拌混合物,最后加入余下的蛋白、淡奶油。

(5) Spoon or pipe the mousse into a serving bowl. Refrigerate for at least 8 hours.

将混合物挤入小碗中,冷藏至少8 h。

第五节 水果甜品的制作工艺

水果香甜多汁,在西餐中尤其是在甜品的制作中被广泛应用。在甜品的制作中,水果的烹调方式是多样的,可生食、烤制、煮制等,一般搭配冰激凌、奶油等。

一、烤苹果配香草冰激凌(baked apple with vanilla ice cream)

 原料 Ingredients

4 large apples	4 个大苹果
1/4 cup of brown sugar	1/4 杯红糖
1 teaspoon of cinnamon	1 茶匙肉桂
1/4 cup of chopped pecans	1/4 杯切碎的山核桃
1/4 cup of chopped raisins	1/4 杯切碎的葡萄干
1 tablespoon of butter	1 汤匙黄油
3/4 cup of boiling water	3/4 杯开水
1 cup(150 mL)	1 杯(150 mL)

 工具 Tools

bowl	搅拌碗
baking pan	烤盘

 制作方法 Method

(1) Preheat oven to 190 ℃. Wash apples. Remove cores to 1/2 inch of the bottom of the apples.

预热烤箱至 190 ℃,洗净苹果,去除苹果果核至苹果底部的 1.27 cm 处。如图 10-2 (彩图 24)所示。

(2) In a small bowl, combine the sugar, cinnamon, raisins, and pecans. Place apples in a 8×8 inch square baking pan. Stuff each apple with this mixture. Top with a dot of butter.

在小碗里混合糖、肉桂、葡萄干碎、核桃碎,制成混合物。将苹果放在 20 cm×20 cm 的烤盘中,将混合物填入苹果中,最后在顶部放一点黄油。如图 10-3(彩图 25)所示。

(3) Add boiling water to the baking pan. Bake for 30-40 minutes, until tender, but not mushy. Remove from the oven and baste the apples for several times with the pan juices.

在烤盘中加入开水,烤制 30~40 min 直至苹果变软,但不要变成糊状。将苹果从烤箱中取出,后用烤盘上的汁水刷苹果的表面。

(4) Serve with vanilla ice cream on the side.

搭配香草冰激凌。

图 10-2

图 10-3

二、糖水炖梨（poached pears）

 原料 Ingredients

500 mL water   500 mL 水
400 g white sugar   400 g 白糖
5 mL vanilla extract   5 mL 香草精华
6 pears   6 个梨子

 工具 Tools

saucepot   炖锅
melon ball cutter   瓜球切割器

 制作方法 Method

（1）Combine the water and sugar in the saucepot, bring to a boil, stirring until the sugar is dissolved.

在炖锅中混合水和糖，将其煮沸，其间不断搅拌直至糖融化。

（2）Remove from heat and add the vanilla.

离火，加入香草精华。

（3）Peel the pears. Cut them in half and remove the cores with a melon ball cutter.

将梨子去皮，对半切，去除果核。

（4）Add the pears to the syrup and simmer very slowly until just tender.

将梨子加入糖浆中，慢炖直至变软。

（5）Let the pears cool in the syrup. When the pears are cool, refrigerate them in the syrup until needed for service.

冷却梨子后放入冰箱冷藏。

三、草莓巧克力(chocolate dipped strawberries)

 原料 Ingredients

10 strawberries, fresh　　　　　　　10 颗新鲜的草莓
50 g white chocolate　　　　　　　　50 g 白巧克力
70 g dark chocolate　　　　　　　　70 g 黑巧克力

 工具 Tools

piping bag　　　　　　　　　　　　裱花袋
microwave　　　　　　　　　　　　微波炉

 制作方法 Method

（1）Wash and dry the strawberries.

将草莓洗净后沥干。

（2）Melt the white chocolate in a microwave.

在微波炉里融化白巧克力。

（3）Dip the strawberry in the white chocolate, holding onto the stem of the strawberry. Give it a quick little twist and shake with your fingers to shake off the excess. Place it on a piece of wax paper to let it dry.

拿住草莓的根部，将草莓浸入白巧克力中，稍微扭一下，然后除去多余的巧克力，将草莓放置在蜡纸上使其变干。

（4）Melt the dark chocolate. Dip the strawberry in at a 45 degree angle from both sides to make the "jacket". Let the excess drip off. Place it on wax paper to dry.

融化黑巧克力，将草莓两边以 45°角浸入巧克力中，后沥去多余的巧克力，将其放置在蜡纸上至变干。如图 10-4（彩图 26）所示。

（5）Place some melted dark chocolate into a piping bag with a very small tip, pipe on buttons and bow tie.

将融化的黑巧克力放入裱花袋中，裱出纽扣和领结。如图 10-5（彩图 27）所示。

（6）Allow to dry and cool. If drying seems to be too slowly, place them in the freezer for about 3-5 minutes.

冷却、干燥。

图 10-4

图 10-5

四、芒果奶油果泥(mango cream puree)

1. 芒果奶油果泥

   原料   Ingredients

1 large, ripe mangoes	1 个大的、成熟的芒果
80 mL custard	80 mL 蛋奶糊
150 mL cream	150 mL 奶油
sliced mango, to serve	切片芒果,装饰

   工具   Tools

food processor　　　　　　　　　　搅拌器

   制作方法   Method

(1) Peel and stone the mangoes and puree the flesh in a food processor. Add the custard and blend to combine.

待芒果去皮、去核后,将果肉放入搅拌器中搅打成泥状,后加入蛋奶糊混合。

(2) Whip the cream until soft peaks form, then gently fold into the mango mixture until just combined, do not overmix.

打发奶油,后轻轻将其加至芒果混合物中,翻拌混合物,不要过度搅拌。

(3) Spoon the mixture into individual glass, refrigerate for at least 1 hour before serving. Serve topped with mango slices.

将混合物放入单独的杯中,冷藏至少 1 h。在其表面放置芒果片进行装饰。

2. 蛋奶糊(custard)

 原料 Ingredients

250 mL milk 250 mL 牛奶
2 teaspoons of cornstarch 2茶匙玉米淀粉
1 tablespoon of white sugar 1汤匙白糖
3 egg yolks 3个蛋黄
1/2 teaspoon of vanilla extract 1/2茶匙香草精华

 制作方法 Method

(1) Mix 2 tablespoons of the milk with the cornstarch in a saucepan. When the cornstarch is dissolved, slowly add the rest of the milk and sugar, and cook over moderate heat until the sauce start to thicken and come to a boil. Remove from heat.

将2汤匙牛奶和玉米淀粉放在一个少司锅里混合,当淀粉融化后,慢慢加入剩下的牛奶和糖,中火搅拌少司直至其变得浓稠,离火。

(2) In a small bowl, beat egg yolks with a fork.

在小碗里,打散蛋黄。

(3) Take a cup of the sauce, and slowly add to the eggs, beat briskly as you pour. Return the egg and sauce mixture to the saucepan, stir into the hot sauce.

将一杯少司慢慢倒入蛋黄中,搅拌均匀,后将混合物倒入少司锅中加热并搅拌。

(4) Bring back to a boil, stir constantly. Remove from heat, and add the vanilla extract.

将混合物煮至沸腾,在煮的过程中偶尔搅拌。最后离火,加入香草精华。

Custard is a variety of culinary preparations based on a cooked mixture of milk or cream and egg yolks. Depending on how much egg or thickener is used, custard may vary in consistency from a thin pouring sauce to a thick pastry cream. Most common custards are used as dessert or dessert sauce and typically include sugar and vanilla.

Custard is usually cooked in a double boiler(bain-marie), or heated very gently in a saucepan on a stove, though custard can also be steamed, baked in the oven with or without a water bath.

蛋奶糊在西餐中被广泛应用,一般用来做甜品或甜品少司汁。根据鸡蛋用量的多少,蛋奶糊的浓度可稀可浓。制作蛋奶糊时,可以在双层蒸锅、少司锅中进行,也可以在烤箱中进行。

五、冰激凌糖水桃子(peach melba)

冰激凌糖水桃子(peach melba)是一道在1893年由法国大厨奥古斯特·埃斯科菲耶 Auguste Escoffier,为澳大利亚女高音歌唱家内莉·梅尔巴创制的甜点。它由放在淡香草味糖浆的沸水中煮过的桃子做成,配上香草冰激凌及覆盆子浓汁。在当代做法中,可用其他水果,如草莓替代桃子。

150 g fresh raspberries	150 g 鲜木莓
1 tablespoon of icing sugar	2 汤匙糖粉
180 g sugar	180 g 糖
1 vanilla bean, split lengthways	1 根香草豆荚,中间分开
2 ripe peaches	2 个成熟的桃子
vanilla ice cream, to serve	香草冰激凌

food processor	搅拌器
strainer	滤器
saucepan	少司锅
slotted spoon	漏勺

(1) Puree the raspberries and icing sugar together in a food processor. Pass through a strainer and discard the seeds. Stir the sugar, vanilla bean and 300 mL water in a saucepan over low heat until the sugar dissolves.

将木莓和糖粉放入搅拌器中打成泥状,后用滤网滤去木莓籽。将糖、香草豆荚和300 mL水放入少司锅中用小火加热直至糖溶解。

(2) Bring the sugar syrup to the boil and add peaches, ensuring they are covered with the syrup. Simmer gently for 5 minutes, or until tender, then remove the peaches with a slotted spoon and carefully remove the skin.

将糖浆煮沸,加入桃子,确保糖浆包裹在桃子上。小火炖 5 min 直至桃肉变软,最后用漏勺将桃子取出来,去皮。

(3) Put a scoop of icecream on a plate, add peach, then spoon the puree over the top.

取一勺冰激凌到盘中,加入桃子,最后淋上木莓果泥。

六、香蕉椰蓉脆饼(banana fritters in coconut batter)

 原料 Ingredients

50 g glutinous rice flour	50 g 糯米粉
30 g desiccated coconut	30 g 椰蓉
30 g sugar	30 g 糖
1/2 tablespoon of sesame seeds	半汤匙芝麻
30 mL coconut milk	30 mL 椰汁
3 bananas	3 根香蕉
oil, for deep-frying	油,油炸用
vanilla ice cream, to serve	香草冰激凌
sesame seeds, toasted, to decorate	芝麻,烤过的,装饰用

 工具 Tools

large bowl	大碗
saucepan	少司锅
slotted spoon	漏勺

 制作方法 Method

(1) Combine the flour, coconut, sugar, sesame seeds, coconut milk and 30 mL water in a large bowl. Whisk to a smooth batter, add more water if the batter is too thick. Set aside to rest for 1 hour.

在一个大碗里混合糯米粉、椰蓉、糖、芝麻、椰汁和 30 mL 水,将它们搅拌成光滑的面糊,如果面糊太浓稠,可以再加入水,后静置 1 h。

(2) Peel the bananas and cut in half lengthways.

将香蕉去皮,纵向对切。

(3) Fill a large heavy-based saucepan one-third full of oil and heat to 180 ℃. Dip each piece of banana into the batter then drop gently into the hot oil. Cook in batches for 4-6 minutes, or until golden brown all over. Remove with a slotted spoon and drain on crumpled paper towels. Serve with ice cream and a sprinkling of toasted sesame seeds.

在一个大锅中加入三成满的油并将其加热至 180 ℃。将每片香蕉蘸上面糊,然后轻轻地放入热油中,炸 4～6 min 直至两面金黄,后用漏勺取出,放置在纸巾上沥干。最后搭配香草冰激凌,表面撒上烤好的芝麻。

第十一章 西式宴会小食

一、西式宴会

宴会是指人们为了社会交往的需要并根据预先计划而举行的群体聚餐活动,具有聚餐式、计划性、规格化和社交性的特征。西式宴会是指摆西式餐台、用西式餐具、吃西式餐菜、按西餐礼仪服务的宴会,主要包括正式宴会、自助宴会(鸡尾酒会和冷餐会)、家宴、便宴等。

二、西式宴会小食

在商务活动的自助餐中,由于受到场地、环境的影响,以及考虑到用餐便利性等因素,经常会提供一些小食给客人食用,例如手指类的小吃(fingerfood)或者开胃小食Canapé(Canapé的法语原意是指以面包为底托做成的手指小食或开胃小食),辅以精致的小块甜品、水果等。虽然在服务上也准备了相应的餐具,但是因为食物取用的便利性,用餐者会经常忽略餐具,而直接通过食物上面的牙签或底托来拿取。

三、宴会小食菜品设计原则

选用工艺流程不太复杂、加工费时少的菜点,选用易于批量制作的菜品。
选用有场地准备和设备加工的菜品。
选用有助于控制菜温、保持外观质量的菜品。
不宜选用口味过于厚重的菜品,例如大蒜等。
选用顾客取食方便且能优雅地进食的菜品。
不宜选用具有过于强烈的刺激性气味的食物。

四、西式宴会小食制作实例

（一）奶酪 & 菠萝 Canapés (cheese & pineapple canapés)

 原料 Ingredients

500 g pineapple pieces	500 g 新鲜的菠萝块
1 bunch of mint	一束新鲜的薄荷
25 g toasted sesame seeds	25 g 烤过的芝麻
250 g pack halloumi	250 g 哈罗米奶酪

哈罗米奶酪是塞浦路斯的传统美食，是一种煎不化的奶酪，它可以像豆腐一样食用，制作简单、方便，可以搭配各种不同的食物。

 工具 Tools

cook knife	西餐刀
mixing bowl	搅拌碗
roasting tray	烤盘
non-stick frying pan	不粘锅

 制作方法 Method

(1) Heat oven to 180 ℃, chop the pineapple into chunks and put them in a roasting tray. Bake for 35-40 minutes until golden, then cover and set aside.

加热烤箱至 180 ℃，将菠萝切成块状放入烤盘，烤制 35～40 min 直至其呈金黄色，后加盖静置。

(2) Chop the mint leaves and mix with the sesame seeds in a dish. Set aside. Cut the halloumi into cubes and char on all sides in a hot non-stick frying pan. Toss the pineapple in the sesame mixture, then thread onto cocktail sticks with a piece of halloumi.

在碗里将切碎的薄荷和芝麻混合，后放置一边。将哈罗米奶酪切成方块状后在不粘锅中煎至四面上色。将菠萝放入芝麻混合物中，然后与一片奶酪一起串在鸡尾酒棒上即可。

(二) 填充鸡尾酒蛋(stuffed cocktail eggs)

原料 Ingredients

6 medium eggs	6 个中等大的鸡蛋
3 tablespoons of yogurt	3 汤匙酸奶
1 teaspoon of mustard	1 茶匙芥末
1 tablespoon of finely chopped parsley	1 汤匙切碎的欧芹
25 g smoked salmon	25 g 烟熏三文鱼
sprigs of fresh dill	新鲜的莳萝
25 g chorizo, skin removed and finely chopped	25 g 西班牙香肠,去皮后切碎

工具 Tools

cook knife	西餐刀
bowl	碗
spoon	勺子
frying pan	炒锅

制作方法 Method

(1) Boil eggs for 7 minutes and put them into iced water to cool. Carefully remove the shells, then cut in half lengthways. Scoop the yolks into a bowl and mash with the yogurt, mustard and parsley. Spoon the mixture back into the eggs.

煮鸡蛋约 7 min,后将其放入冰水中冷却,剥去蛋壳后对半切开,将蛋黄挖到碗里,与酸奶、芥末、欧芹一起捣烂,最后将混合物填至鸡蛋中。

(2) For the salmon eggs, top each with a strip of salmon and add some fresh dill.

制作三文鱼鸡蛋时,在鸡蛋上放置一片三文鱼和新鲜的莳萝。

(3) For the chorizo version, fry the chorizo gently in a pan until it is crisp. Scatter over the eggs when cool. Keep chilled until ready to serve.

制作西班牙香肠鸡蛋时,将香肠煎至松脆,待其放凉后撒在鸡蛋上,冷藏。

(三)生菜配芒果、虾仁沙拉(lettuce bowl with mango and prawn)

 原料 Ingredients

3 little lettuces, broken into individual leaves	3棵小生菜,分成单片叶子
1 ripe mango	1个成熟的芒果
1 finely chopped red onion	1个切碎的紫洋葱
1 chopped red chilli	1个切碎的红辣椒
lime juice	青柠汁
4 tablespoons of olive oil	4汤匙橄榄油
225 g pack cooked, peeled tiger prawn	225 g煮熟、去皮的老虎虾
a handful of chopped coriander	一把切碎的欧芹

 工具 Tools

cook knife	西餐刀
bowl	碗
spoon	勺子

 制作方法 Method

(1) Break the lettuces into individual leaves. Peel and dice the mango and mix with the finely chopped red onion, chopped mild red chilli, the lime juice and the olive oil. Mix well and set aside at room temperature, covered with cling film.

将生菜分成单片的叶子。将芒果去皮、切丁,后与切碎的紫洋葱、不辣的红辣椒、青柠汁、橄榄油混合均匀,封上保鲜膜,在室温下放置。

(2) Just before serving, take peeled tiger prawns and cut them in half horizontally. Mix into the mango mixture with a handful of chopped coriander. To serve, spoon the mixture on the little leaves and arrange on a large plate.

食用前,将老虎虾去皮、水平切开,后与芒果混合物、切碎的欧芹混合。最后用勺子将混合物放在生菜叶片上,放置在盘子里。

(四)红薯堆(sweet potato stackers)

 原料 Ingredients

2 large sweet potatoes (peeled and each cut into 8 chunky pieces)	2 个大红薯(去皮、切成大块)
1 tablespoon of olive oil	1 汤匙橄榄油
2 tablespoons of mayonnaise	2 汤匙蛋黄酱
lemon juice	柠檬汁
8 slices of prosciutto, halved	8 片意大利熏火腿,切成两半
few watercress sprigs	豆瓣菜适量

 工具 Tools

baking sheet	烤盘
platter	浅盘

 制作方法 Method

(1) Heat oven to 200 ℃, then toss the potato chunks with the olive oil and some seasoning on a baking sheet. Roast for 20-30 minutes until the surface is golden and crisp, then leave them to cool.

预热烤箱至 200 ℃,将红薯块、橄榄油、调味料一起放入烤盘,烤制 20～30 min 直至其表面金黄、酥脆,放凉。

(2) To serve, mix the mayonnaise with the lemon juice. Pile a scrunched up piece of ham on each potato, then top with a blob of the lemony mayo. Arrange them on a platter with the watercress, then serve.

将蛋黄酱、柠檬汁混合,再将折叠的火腿片放置在红薯上,最后在上面放柠檬蛋黄酱,将其放置在浅盘中,并用豆瓣菜装饰。

(五)山羊奶酪、培根酿甜椒(goat cheese and bacon stuffed peppers)

 原料 Ingredients

300 g mini sweet pepper	300 g 小甜椒
2 tablespoons of oil	2 汤匙油
salt and peppercorn	盐、胡椒

250 g soft goat cheese

3 pieces of bacon, cooked and crumbled

1 tablespoon of fresh chives, sliced

250 g 山羊奶酪

3 片培根,熟的、碎的

1 汤匙新鲜的青葱,切碎的

工具 Tools

baking sheet 烤盘

small bowl 小碗

制作方法 Method

(1) Preheat oven to 300 ℃.

预热烤箱至 300 ℃。

(2) Line baking sheet with foil and set aside.

在烤盘上铺上锡纸。

(3) Slice the tops of the mini sweet peppers off and cut them in half and remove seeds. Lay halved mini sweet peppers in a single layer on baking sheet and drizzle with oil.

将甜椒的顶部去掉后将其切成两半,去除籽,将切好的甜椒平铺在烤盘上,撒上油。

(4) In a small bowl, mix salt, peppercorn, and goat cheese until smooth. Fill each mini sweet pepper with the cheese.

在一个小碗里混合盐、胡椒和山羊奶酪,搅拌混合物直至均匀,后将其填入甜椒中。

(5) Bake for 15 minutes until cheese is nice and brown. Remove from oven and top with bacon and chives.

烤制 15 min 直至奶酪上色,后将其从烤箱中取出,在表面撒上培根、青葱。

(六)意大利烤面包片(bruschetta on a grilled baguette)

原料 Ingredients

200 g cherry tomatoes, sliced

2 cloves of garlic, minced

8 basil leaves, chopped

2 tablespoons of balsamic vinegar

3 tablespoons of olive oil

salt and peppercorn

1/2 baguette

200 g 樱桃番茄,切片

2 瓣大蒜,切碎的

8 片罗勒叶,切碎的

2 汤匙意大利香脂醋

3 汤匙橄榄油

盐和胡椒

半根法棍面包

工具 Tools

skillet	煎锅
small bowl	小碗

 制作方法 Method

(1) Heat 1 tablespoon of olive oil over medium heat. Add minced garlic and saute for about 1 minute. Remove from pan to cool.

先用中火加热 1 汤匙橄榄油,再加入切碎的大蒜炒 1 min,后离火。

(2) Add cherry tomatoes, sauteed garlic, basil leaves, balsamic vinegar, olive oil, salt and peppercorn in a bowl and gently combine them. Set aside.

在碗里混合樱桃番茄、炒香的大蒜、罗勒叶、意大利香脂醋、橄榄油、盐、胡椒后,再将混合物静置。

(3) Slice baguette, heat 1 tablespoon of olive oil in a skillet over medium heat. Add baguette slices and grill on each side until nice and brown for about 3 minutes per side.

将法棍面包切片,在锅中放入 1 汤匙橄榄油,并用中火加热,将切好的法棍面包放入锅中煎至上色。

(4) Top grilled baguette with bruschetta and serve.

在法棍面包上放上番茄混合物。

(七)番茄马苏里拉串烧(tomato Mozzarella skewers)

 原料 Ingredients

cherry tomatoes	樱桃番茄
several leaves of fresh basil	若干片新鲜的罗勒叶
fresh Mozzarella cheese, cut into small squares	新鲜的马苏里拉奶酪,切成小块
salt and peppercorn	盐、胡椒
balsamic vinegar	意大利香脂醋
tooth picks or skewers	装饰签或串肉杆

 制作方法 Method

(1) Thread tooth picks or skewers with a cherry tomato, a basil leaf, and Mozzarella cheese. Add salt and peppercorn and drizzle with balsamic vinegar.

用装饰签或串肉杆将樱桃番茄、罗勒叶、马苏里拉奶酪穿成串,并将其撒上盐、胡椒、意大利香脂醋。

(2) Serve cold or at room temperature.

冷食或室温食用。

(八)烤鸡肉配奶油牛油果蘸汁(baked chicken tenders with creamy avocado dipping sauce)

1. 烤鸡肉

 原料 Ingredients

400 g boneless chicken breast, sliced	400 g 鸡胸肉,切片
100 g whole wheat flour	100 g 全麦面粉
1 egg, beaten	1个鸡蛋,打散
100 g whole wheat bread crumbs	100 g 全麦面包屑
salt and peppercorn	盐、胡椒

 制作方法 Method

(1) Preheat oven to 350 ℃. Spray a baking sheet with cooking spray and set aside.

预热烤箱至350 ℃,在烤盘上涂上不粘涂层喷雾.

(2) Add flour, beaten egg, and bread crumbs in three separate bowls. Add salt and peppercorn to the bowl with bread crumbs.

在三个单独的碗里分别加入面粉、打散的鸡蛋、面包屑,在装有面包屑的碗里加入盐、胡椒。

(3) Dip the tender chicken in the flour until lightly coated. Then dip them into egg and then into the bread crumbs.

将鸡肉依次拍上面粉,拖上蛋液,再蘸上面包屑。

(4) Place chicken in a single layer on the baking sheet and bake for about 15-20 minutes depending on the thickness of the chicken.

将单层鸡肉放入烤盘中,烤制15~20 min 即可。

2. 奶油牛油果蘸汁

 原料 Ingredients

1 avocado, peeled and pitted	一个牛油果,去皮、去核
1/2 lime, juiced	半个青柠,榨汁

salt	盐适量
1 tablespoon of yogurt	1 汤匙酸奶
100 mL water	100 mL 水

 制作方法 Method

Blend the avocado, lime juice, salt, and yogurt in a food processor and mix them until smooth.

在搅拌器中混合牛油果、青柠汁、盐、酸奶,搅拌混合物直至均匀。

主要参考文献

[1] 李祥睿.西餐工艺[M].北京:中国纺织出版社,2008.
[2] 郭亚东.西餐工艺[M].北京:中国轻工业出版社,2006.
[3] 周海霞,邹宇航.西餐制作[M].北京:科学出版社,2014.
[4] 牛铁柱,林粤,周桂禄.西餐烹调工艺与实训[M].北京:科学出版社,2013.
[5] 陆理民.西餐工艺与实训[M].北京:中国旅游出版社,2013.
[6] Wayne Gisslen. Professional Cooking [M]. Hoboken:Wiley,2010.
[7] 谢军.从西餐看中国餐饮业现状与发展[J].商,2015,(24):270-271.
[8] 陈尘.浅谈中国饮食文化对西餐的影响[J].成功(教育版),2011,(14):278.
[9] 陈忠明.西餐的传入及其对中国餐饮业的影响[J].扬州大学烹饪学报,2003,20(2):49-53.
[10] 邵万宽.近现代西餐烹饪技艺在中国的渗透与发展[J].南宁职业技术学院学报,2017,22(3):1-5.
[11] 许二凤,赵霞,林静.西餐在中国的发展研究[J],四川烹饪高等专科学校学报,2012,(1):48-51.

彩 图

彩图 1

彩图 2

彩图 3

彩图 4

彩图 5

彩图 6

彩图 7

彩图 8

彩图 9

彩图 10

彩图 11

彩图 12

彩图 13

彩图 14

彩图 15

彩图 16

彩图 17

彩图 18

彩图 19

彩图 20

彩图 21

彩图 22

彩图

彩图 23

彩图 24

彩图 25

彩图 26

彩图 27